Lecture Notes in Artificial Intelligence 2246

Subseries of Lecture Notes in Computer Science
Edited by J. G. Carbonell and J. Siekmann

Lecture Notes in Computer Science

Edited by G. Goos, J. Hartmanis, and J. van Leeuwen

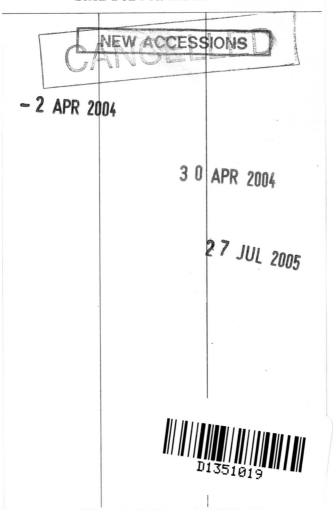

Springer

Berlin
Heidelberg
New York
Barcelona
Hong Kong
London
Milan
Paris
Tokyo

Rino Falcone   Munindar Singh
Yao-Hua Tan (Eds.)

Trust in Cyber-societies

Integrating the Human and Artificial Perspectives

 Springer

Series Editors

Jaime G. Carbonell, Carnegie Mellon University, Pittsburgh, PA, USA
Jörg Siekmann, University of Saarland, Saarbrücken, Germany

Volume Editors

Rino Falcone
National Research Council, Institute for Cognitive Science and Technology
Group of Artificial Intelligence, Cognitive and Interacting Modelling
Viale Marx 15, 00137 Rome, Italy
E-mail: falcone@ip.rm.cnr.it

Munindar Singh
North Carolina State University, Department of Computer Science
940 Main Campus Drive, Suite 110, Raleigh, NC 27606, USA
E-mail: mpsingh@eos.ncsu.edu

Yao-Hua Tan
Free University Amsterdam, Department of Economics and Business Administration
De Boelelaan 1105, 1081 HV Amsterdam, The Netherlands
E-mail: ytan@feweb.vu.nl

Cataloging-in-Publication Data applied for

Die Deutsche Bibliothek - CIP-Einheitsaufnahme

Trust in cyber societies : integrating the human and artificial perspectives
/ Rino Falcone ... (ed.). - Berlin ; Heidelberg ; New York ; Barcelona ;
Hong Kong ; London ; Milan ; Paris ; Tokyo : Springer, 2001
  (Lecture notes in computer science ; Vol. 2246 : Lecture notes in
artificial intelligence)
  ISBN 3-540-43069-5

CR Subject Classification (1998): I.2, H.5.3, K.4

ISSN 0302-9743
ISBN 3-540-43069-5 Springer-Verlag Berlin Heidelberg New York

Springer-Verlag Berlin Heidelberg New York
a member of BertelsmannSpringer Science+Business Media GmbH

http://www.springer.de

© Springer-Verlag Berlin Heidelberg 2001
Printed in Germany

Typesetting: Camera-ready by author, data conversion by Steingräber Satztechnik GmbH, Heidelberg
Printed on acid-free paper   SPIN: 10845915   06/3142   5 4 3 2 1 0

Preface

This book is the result of the workshop "Deception, Fraud, and Trust in Agent Societies", held in Barcelona on June 4, 2000 as part of the Autonomous Agents 2000 Conference, and organized by Rino Falcone, Munindar Singh, and Yao-Hua Tan. The aim of the workshop was to bring together researchers from different fields (Artificial Intelligence, Multi-Agent Systems, Cognitive Science, Game Theory, and Social and Organizational Sciences) that could contribute to a better understanding of trust and deception in agent societies. The workshop scope included theoretical results as well as their applications in human-computer interaction and electronic commerce.

This book includes the revised and extended versions of the works presented at the workshop, incorporating many points that emerged in our discussions, as well as invited papers from experts in the field, which in our view allows a complete coverage of all relevant issues. We gratefully acknowledge the financial support from the Italian National Research Council - Institute for Cognitive Science and Technology and the ALFEBIITE European Project, contract number IST-1999-10298.

We would like to express our gratitude to Cristiano Castelfranchi for his stimulating and valuable comments and suggestions both for the organization of the workshop and for the preparation of this book.

We would also like to thank Andreas Birk, Valeria Carofiglio, Cristiano Castelfranchi, Rosaria Conte, Robert Demolombe, Fiorella de Rosis, Aldo Franco Dragoni, Maria Miceli, Samuel Santosa, Onn Shehory, Chris Snijders, Von-Won Soo, Walter Thoen, Gerd Wagner, and Eric Yu for their important help in reviewing the papers of this book.

Finally, we are indebited to Rosamaria Romano for her support in editing this book, and to Albano Leoni for his dedicated secretarial support.

October 2001

Rino Falcone
Munindar Singh
Yao-Hua Tan

Sponsoring Institutions

Italian National Research Council
Institute for Cognitive Science and Technology

ALFEBIITE European Project, contract number IST-1999-10298

Table of Contents

Introduction:
Bringing Together Humans and Artificial Agents in Cyber-societies: A New Field of Trust Research

Rino Falcone[1], Munindar Singh[2], and Yao-Hua Tan[3]

[1] National Research Council Institute for Cognitive Science and Technology
Group of "Artificial Intelligence, Cognitive and Iteracting Modelling"
Viale Marx, 15
00137 Roma
Italy
falcone@ip.rm.cnr.it
[2] Department of Computer Science
North Carolina State University
USA
mpsingh@eos.ncsu.edu
[3] Department of Economics and Business Administration
Free University Amsterdam
De Boelelaan 1105
1081 HV Amsterdam
The Netherlands
ytan@feweb.vu.nl

Humans have learned to cooperate in many ways and environments; on different tasks; and for achieving different goals. Collaboration and cooperation in their more general sense (and in particular: negotiation, exchange, help, delegation, adoption, and so on) are an important characteristic — or better, one of the foundational aspects — of human societies.

In the evolution of cooperative models diverse constructs of various kinds (e.g. purely interactional, technical-legal, organizational, socio-cognitive, etc.), have been opportunely introduced or spontaneously emerged to support decision making in collaborative situations.

The new scenario we are going to meet in the third millennium transfigures the old frame of reference, in that we have to consider: new channels and infrastructures (i.e. Internet); new artificial entities for cooperating with (artificial or software agents); new modalities of interaction (suggested/imposed by both the new channels and the new entities). Thus, it will be necessary to update the traditional supporting decision making constructs. This effort will be necessary especially to develop the new cyber-societies in such a way as not to miss some of the important cooperative characteristics which are so relevant in human societies.

Trust, in the general frame above described, might be considered as a socio-cognitive construct of main importance. In particular, trust building is now recognized as a key factor for using and developing the new interactional paradigm.

Trust should not be reduced to mere security. The latter can be useful to protect from the intrusion of an unknown agent, to guarantee an agent of the

R. Falcone, M. Singh, and Y.-H. Tan (Eds.): Trust in Cyber-societies, LNAI 2246, pp. 1–7, 2001.
© Springer-Verlag Berlin Heidelberg 2001

identity of its partner, to identify the sender of a message (for example by verifying the origin of a received message; by verifying that a received message has not been modified in transit; by preventing that an agent who sent a message might be able to deny later that it sent the message [1]). With sophisticated cryptographic techniques it is possible to give some solution to these security problems.

However, more complex is the issue of trust, that must give us tools for acting in a world that is in principle insecure (that cannot be considered 100% secure), where we have to make the decision to rely on someone in risky situations. Consider the variety of cases in which it is necessary or useful to interact with agents whose identity, history or relationships are unknown, and/or it is only possible to make uncertain predictions on their future behaviours.

Probably the most acute example of the need of trust building is in Electronic Commerce, but trust building is also essential in other domains of Multi Agent Systems and Agent Theory such as Agent Modelling, Human-Computer Interaction, Computer Supported Cooperative Work, Mixed Initiative and Adjustable Autonomy, and Ubiquitous Computing.

In fact, various different kinds of trust should be modelled, designed, and implemented:

- trust in the environment and in the infrastructure (the socio-technical system);
- trust in personal agents and in mediating agents;
- trust in potential partners;
- trust in information sources;
- trust in warrantors and authorities.

Part of these different kinds of trust have a complementary relations with each other, that is, the final trust in a given system/process can be the result of various trust attributions to the different components. an exemplary case is one's trust in an agent that must achieve a task (and more specifically in its capabilities for realizing that task) as different from one's trust in the environment (hostile versus friendly) where that agent operates, or again as different from one's trust in a possible third party, (arbitrator, mediator, normative systems, conventions, etc.) able to influence or constraint the trustee and presenting a guaranty for the trustor [2,3].

Therefore, the "sufficient" trust value of one single component cannot be established before evaluating the value of the other components. In this regard, it is very interesting to characterize the relationships between trust and (partial) control [4].

It is important to underline how trust is in general oriented towards not directly observable properties; it is, in fact, based on the ability to predict these properties and to rely or not on them. Thus, it is quite complex to assess the real trustworthiness of an agent, system, or process, not only because -as we have seen- there are many different components that contribute to this trustworthiness, but also because the latter is not directly observable (see [5] about signs of trust). The important thing is the perceived trustworthiness that is, in its turn, the result of different modalities of the trustor's reasoning about: direct experience; categorization; inference; communicated reputation.

To summarize, we are moving towards completely new modalities, infrastructures, and actors of the cooperative paradigm in which human factors will be integrated and supported by artificial technologies, and in which it will be difficult to assess whether the cooperating partners are humans or artificial agents, or a mixed community. This endeavour will be more successful if the socio-psychological factors will continue to play a fundamental role, which implies incorporating, adapting and developing them within these new technologies. For this reason, an important role is given by those studies and researches about models, implementations and experiments that try to fill this gap. In this scenario, trust is among the most relevant factors of the new cooperative models and this book addresses many of the key issues in contemporary research and theorizing on trust in cyber-societies. In the next session, the main themes covered in each chapter will be outlined.

Outline of the book chapters

When an agent has to make a decision about whether to trust or not another agent in the perspective of a cooperative relationship, he must weight the opportunities given by the positive results of a successful trust (benefits of trust) against the risks that his trust might be exploited and betrayed: this problem is known as the trust dilemma. The trust dilemma is the direct consequence of uncertainty, in this case of the intrinsic social uncertainty.

In his paper Trust Rules for Trust Dilemmas: How Decision Makers Think and Act in the Shadow of Doubt, Roderick Kramer — starting from the assumption that individuals possess various kinds of rules to use when trying to make sense of what to do in trust dilemma situations (social auditor model) — investigates the comparative efficacy of different decision rules regarding trust in a simulated trust dilemma. In particular he analyzes and compares different classes of trustor (characterized by different trust rules), such as, social optimists (that possess benign views about other agents) and vigilants (that worry about the possibility of being exploited). The results of these simulations are really interesting and not trivially predictable.

The problem of the definition of trust and its corresponding meaning and component features is relevant not only to a deeper comprehension of the trust phenomenon but also for modeling, projecting, developing, and implementing trust in the future networks for cooperation. The work by Harrison McKnight and Norman Chervany on Trust and Distrust Definitions: One Bite at a Time develops a conceptual typology of trust and distrust concepts, and defines, as subsets of the high level concepts, measurable constructs for empirical research. The authors explore different approaches to trust from different disciplines (Psychology, Sociology, Economy, Management and Communication, Political Science, etc.) and extract four high level categories for the trustee: benevolence, integrity, competence, and predictability. They connect trust with power and control and also suggest that trust and distrust are separate constructs that may co-exist.

Trust is clearly a dynamic phenomenon, which changes in time on the basis of the trustor experience: this is a quite intuitive assumption, even if it is not so trivial to introduce it in a computational model. However, there is also another way to consider the dynamics of trust: that is to look at its dynamic self-reproductive power.

Rino Falcone and Cristiano Castelfranchi approach this particular case of trust dynamics in their work "The socio-cognitive dynamics of trust: Does trust create trust".

They address how the trustworthiness of the trustee might be influenced and modified by the trustor's act of trust; how this change in trustworthiness can be anticipated by the trustor and taken into account in a new attribution of trustee's trustworthiness; how the trustor's trust in the trustee modifies the trustee's trust in the trustor; and finally, how diffuse trust diffuses trust (in several ways and for various reasons).

Agents can receive information not just from their direct experience with the world (perception) but also from other information sources. In order to use such information, they should (explicitly or implicitly) trust them. In fact, information credibility is a function of its sources, depending on both the credibility of the sources and the convergence/divergence rate among the different sources. Suzanne Barber and Joon Woo Kim in their paper Belief Revision Process based on Trust: Agents Evaluating Reputation of Information Sources, propose a multi-agent belief revision algorithm that utilizes knowledge about the reliability or reputation of information sources. Trust and reputation are viewed from an economical perspective, where reputation is considered as an asset, something that provides a monetary flow to its owner. This enforces the individual agent's "soft security" rather than its hard security (infrastructure level security such as secure communication with public keys or authenticated name services, etc.). The belief revision algorithm provides efficient numerical treatment of inherent uncertainty due to the nature of practical application domains, and combining existing techniques such as Demster-Shafer theory of evidence and Bayesian belief networks.

In their work Adaptive Trust and Co-operation: An Agent-based Simulation Approach, Bart Nooteboom, Tomas Klos, and Renè Jorna discuss the Transaction Cost Economics (TCE) methodology arguing that it is just partially adequate to model inter-firm relations (even if it introduces useful insights). Conversely, the authors use the Agent-based Computational Economics (ACE) methodology through which they can model how cooperation, trust and loyalty emerge and shift adaptively as relations evolve in a context of multiple, interacting agents. In general, ACE is a good methodology to deal with "complex interrelated structures" of 'processes of interaction in which future decisions are adapted to past experiences'. The authors specify, within the overall framework of ACE, a process of boundedly rational adaptation, based on a mutual evaluation of transaction partners that takes into account trust and profits. Trust is central in this work because while TCE does not incorporate trust, ACE under-lines the role of adaptation in the light of experience, which seems relevant to trust.

Mark Witkowsky, Alexander Artikis and Jeremy Pitt in their paper Experiments in Building Experential Trust in a Society of Objective-trust Based Agents consider the trust based trading relationship as implying three components: reputation, belief based trust, and objective trust. In particular, they develop the so-called objective trust, that is trust of, or between, agents based on actual experiences between those agents. In particular, in the authors' view, agents select who they will trade with primarily on the basis of a trust measure built on past experiences of trading with those individuals. The paper addresses three main questions: What happens when agents, who rank experiential trust and trustworthiness highly, form into trading societies? Does a trust relationship established between agents over a period of time lead to loyalty between those agents when trading conditions become difficult? Trust, however it is evaluated, is personal; is it in an agent's interest to appear trustworthy in some cases, and not care in others?

Trust building can be interestingly viewed as a phenomenon with a double temporal scale. On the one hand, natural evolution permits the emergence of a general predisposition to trust. On the other hand, individual social interactions in a specific environment characterize the actual trust developed through some learning mechanism. A model that permits to clearly identify and distinguish between the parameters influenced by either the first or the second type of trust would be of great importance. Andreas Birk in his paper Learning to Trust, presents a research in which cooperation and trust can develop together through social interactions and a suited learning mechanism. He also discusses evolutionary game-theory as a powerful tool to investigate the development of complex relations between individuals such as the emergence of cooperation and trust. This research uses the so-called multiple-hypotheses approach to successfully develop cooperation and trust simultaneously in scenarios modelled by a continuous-case N-player prisoner's dilemma.

Rajatish Mukherjee, Bikramjit Banerjee, and Sandip Sen in their work Learning Mutual Trust consider the problem of learning techniques in multiagent systems. In these systems — in which there are no isolated single agents but many agents cooperating and/or competing with each other — the standard reinforcement learning techniques are not guaranteed to converge. In particular, the desired convergence in multi-agent systems is toward an equilibrium strategy-profile (collection of strategies of the agents) rather than optimal strategies for an individual agent. For example, Nash Equilibrium does not guarantee that the agents will obtain the best possible payoffs. In this work it is studied if agents can learn to converge to a more desirable pareto-optimal solution than a Nash-Equilibrium. For obtaining this different solution the authors introduce mutual trust as a mechanism that can enable the players to stick to non-Nash Equilibrium action combination (that may yield better payoffs for both agents). The authors address mutual trust because such desirable non-myopic action choices must be preferred by both agents.

Electronic certification is one of the most widespread tools for establishing security in electronic commerce. An electronic certificate is an electronic document signed by some issuer which contains some claims about a subject. The subject is represented by a public key and it should keep the private key that corresponds to the public one to prove that it is really that subject. Although security is just a facet of trust, it is very relevant also for supporting trust itself. Yosi Mass and Onn Shehory in their paper Distributed Trust in Open Multi-Agent Systems, adopt an original certificate-based approach where, in particular, the identity disclosure is not required for trust establishment; and there is no need for a centralized certification mechanism (or globally known trusted third parties). The process of trust establishment, in this work, is performed in a fully distributed manner, where any party or agent may be a certificate issuer, and it is not required that certificate issuers be known in advance. In other words, when requested, an agent that issues certificates provides sufficient certificates from other issuers to be considered a trusted authority according to the policy of the requesting party.

Eric Yu and Lin Liu in their paper Modelling Trust for System Design Using the i* Strategic Actors Framework, introduce the concept of a softgoal to model quality attributes for which there are no a priori, clear-cut criteria for satisfaction, but are judged by actors as being sufficiently met (satisfied) on a case-by-case basis. Trustworthiness is treated as a softgoal to be satisfied from the viewpoint of each stakeholder. The i* framework was developed for modelling intentional relationships among strategic actors. Actors have freedom of action, but operate within a network of social relationships. Specifically, they depend on each other for goals to be achieved, tasks to be performed, and resources to be furnished. In the i* approach, trust is not treated as a distinguished concept with special semantics, but as a non-functional requirement that arises in multi-agent collaborative configurations. In these configurations agents depend on each other in order to succeed. Some configurations may require more trust than others. For example, the financial institution and the customer in a stored valued card system are taking more risks than the phone company and customer in a prepaid phone card system.

References

[1] Q. He, K. Sycara, and Z. Su, (2001), *Security infrastructure for software agent society*, in C. Castelfranchi and Y. Tan (Eds), Trust and Deception in Virtual Societies, Kluwer Academic Publishers, pp.139-156.

[2] Castelfranchi C., Falcone R., (1998) *Principles of trust for MAS: cognitive anatomy, social importance, and quantification*, Proceedings of the International Conference on Multi-Agent Systems (ICMAS'98), Paris, July, pp.72-79.

[3] R. Falcone and C. Castelfranchi, (2001), Social Trust: A Cognitive Approach, in C. Castelfranchi and Y. Tan (Eds), Trust and Deception in Virtual Societies, Kluwer Academic Publishers, pp.55-90.

[4] Castelfranchi C., Falcone R., (2000), *Trust and Control: A Dialectic Link, Applied Artificial Intelligence journal*, Special Issue on Trust in Agents Part1, Castelfranchi C., Falcone R., Firozabadi B., Tan Y. (Editors), Taylor and Francis 14 (8), pp. 799-823.

[5] M. Bacharach and D. Gambetta, (2001), *Trust as Type Detection*, in C. Castelfranchi and Y. Tan (Eds), Trust and Deception in Virtual Societies, Kluwer Academic Publishers, pp.1-26.

Trust Rules for Trust Dilemmas: How Decision Makers Think and Act in the Shadow of Doubt

Roderick M. Kramer

Graduate School of Business
Stanford University
`kramer_roderick@gsb.stanford.edu`

Abstract. Trust has long been recognized as an important antecedent of cooperative behaviour. For example, trust facilitates the productive exchange of informaion in collaborative relationships. Central to the decision to trust another individual in such situations is the <u>trust dilemma</u>: even though recognizing the benefits of trust, individuals recognize also the prospect that their trust might be betrayed. Thus, they must decide how much trust (or distrust) is warranted. At a psychological level, this trust dilemma is animated by social uncertainty (uncertainty regarding the other party's motives, intentions and actions). Using a computer simulation methodology, this paper investigates the comparative efficacy of different decision rules regarding trust in a simulated trust dilemma. The results demonstrate that attributional generosity (operationalized as giving the other party the benefit of the doubt) facilitates the development and maintance of more cooperative relationships when social uncertainty is present.

1 Trust Rules for Trust Dilemmas: How Decision Makers Think and Act in the Shadow of Doubt

"In a broad perspective, rules consist of explicit or implicit norms, regulations, and expectations that regulate the behavior of individuals and interactions among them. They are a basic reality of individual and social life; individual and collective actions are organized by rules, and social relations are regulated by rules."
– March, Schulz, & Zhou (2001, p. 5).

Imagine the following hypothetical vignette. Two researchers working on a problem of mutual interest decide to enter into a scientific collaboration. They each agree to share all of the ideas and empirical findings emerging from their individual research laboratories with the aim of joint publication. Imagine further that they are working on a problem of considerable scientific and social importance (e.g., finding a vaccine for the AIDS virus). The prospects for a fruitful collaboration between the researchers appear excellent because each brings to the table distinctive but complementary competencies: One is a particularly gifted theorist, the other an unusually skilled experimentalist.

R. Falcone, M. Singh, and Y.-H. Tan (Eds.): Trust in Cyber-societies, LNAI 2246, pp. 9–26, 2001.
© Springer-Verlag Berlin Heidelberg 2001

From the standpoint of each researcher, therefore, the collaboration represents a unique and possible invaluable opportunity. Given the importance of the problem, it is reasonable to expect that any individual or individuals credited with eventually solving the problem under investigation will receive considerable acclaim, possibly even a Nobel Prize. It is an opportunity, however, that is obviously attended by some vulnerability. Neither collaborator knows a great deal about the other. Both have worked mostly alone in the past. Thus, there is little readily available information either can use to assess the trustworthiness of the other as a collaborator. Moreover, the costs of misplaced trust are potentially quite steep if one of the researchers "defects" from the collaboration in the end stages, he or she may be able to garner the lion's share of the credit and acclaim for solving the problem.

The contours of this vignette, although hypothetical, are hardly unusual. In the highly competitive world of science, the opportunities for sharing effort and insight are substantial, as are the vulnerabilities, as Watson's surprisingly candid account of the discovery of the molecular configuration of DNA, recounted in *Double Helix*, made quite evident to the general public. Such situations are but one example of a broad class of decision problems known as <u>trust dilemmas</u>. In a trust dilemma, a social decision makers hopes to reap some perceived benefit from engaging in trusting behavior with another social decision maker. Pursuit of the opportunity, however, exposes the decision makers to the prospect that his or her trust might be exploited and betrayed. This conjunction of opportunity and vulnerability are the <u>sine qua non</u> of a trust dilemma. Because of our dependence on, and interdependence with, other social decision makers, trust dilemmas are an inescapable feature of social and organizational life.

At a psychological level, trust dilemmas are animated, of course, by uncertainty about the trustworthiness of the other decision maker(s) with whom the individual is interdependent. It is not knowing the other's character, motives, intentions and actions that make trust desirable but risky. Social uncertainty of this sort, indeed, is fundamental to the problem of trust. As Gambetta, (1988) aptly observed in an early and influential analysis of this problem, "The condition of ignorance or uncertainty about other people's behavior is central to the notion of trust. It is related to the limits of our capacity ever to achieve a full knowledge of others, their motives, and their responses to endogenous as well as exogenous changes" (p. 218).

How do decision makers in trust dilemma situations cope with such uncertainty? How do they decide, for example, how much trust or distrust is appropriate when dealing with another decision makers about whom they possess incomplete social information? To what extent do they make fairly positive presumptions about others and to what extent are they likely to entertain fairly negative or pessimistic expectations? These are the fundamental questions I engage in this paper. To explore them, I examine some relationships between different 'trust rules' and their consequences within the context of 'noisy' (socially uncertain) decision contexts.

I approach these questions from the standpoint of one perspective on how social decision makers make sense of and respond to the social uncertainty intrinsic to trust dilemmas. I characterize this perspective as the social auditor model. After providing a brief overview of this model, I present some evidence for the model and then elaborate on some of its implications.

2 The Social Auditor Model

In almost every domain of life, people rely on various kinds of rules to help them assess problems and make decisions. For example, chess players have rules for interpreting complex endgames and for prescribing the best moves to make in those situations. Physicians have rules for distinguishing between patient complaints that require further investigation versus mere reassurance that nothing serious is the matter. And taxi drivers have rules for deciding who to let into their cab late at night and whom they should pass by. Similarly, social decision makers possess rules for helping them form judgments and reach decisions in interdependence dilemmas as well (see Messick & Kramer, 2001).

According to the social auditor model, individuals possess various kinds of rules to use when trying to make sense of what to do in trust dilemma situations (see Kramer, 1996 for a fuller elaboration of this framework). These rules include interpretation rules (e.g., rules that help us categorize a given trust dilemma and prescribe the sort of evidence we should look for when trying to assess another decision maker's trustworthiness) and action rules (rules about what behaviors we should engage in when responding to those interpretations). To return to the example with which I opened this chapter, we can imagine an individual involved in a potentially fruitful cooperative relationship who has to decide how much to cooperate with the other, given some uncertainty regarding their trustworthiness.

According to the social auditor model, people navigate through trust dilemmas using their mental models of the world. These mental models include their social representations, which encompass everything they believe about other people, including all of their trust-related beliefs and expectations, their self representations (e.g, their beliefs about their own sophistication and competence at judging others' trustworthiness), and their situational taxonomies (e.g., their beliefs about the various kinds of social situations they are likely to encounter in their social lives). These mental models are used to help people interpret the trust dilemmas they confront. On the basis of these interpretations, they make choices.

These choices can be conceptualized as the action rules individuals use when responding to trust dilemmas. Action rules thus represent decision makers' beliefs about the "codes of [prudent] conduct" they should appeal to and employ when trying to act in trust dilemma situations. Interpretation and action rules are viewed in this framework as intendedly adaptive orientations. In other words, in using them, decision makers think they will help them 1) reap the benefits of trust when one is dealing with a trustworthy other and 2) minimize the costs of misplaced trust when one is interacting with an untrustworthy other.

According to the social auditor model, decision makers also monitor the consequences of rule use in trust dilemma situations. In other words, they pay attention to what happens to them after they have employed a given rule when interacting with a specific other. During this post-decision "auditing" process, they attempt to discern, for example, whether the amount of trust they have displayed toward the other is prudent or imprudent. Was too much trust afforded or too little?

On the basis of the conclusions they reach, they are likely to change their subsequent behavior when interacting further with the individual. The results of the post-decision auditing process then are assumed to inform (i.e., validate or invalidate) one's mental model of the dilemma, resulting in possible modification of the rule system invoked, leading to a change in rule use, etc. The model thus posits a cyclic, adaptive learning system of the sort described by March and his colleagues (March, Schulz, & Zhou, 2000).

If we approach the problem of the "correct" or "optimal" level of trust or distrust to manifest when dealing with other people in a decision making ecology in which social uncertainty is present, we can consider what properties of an interpretation-action rule system are likely to produce the best results. For instance, does it makes sense to be a bit "paranoid" when dealing with others we don't know well, making fairly pessimistic or bleak assumptions about their willingness to behave in a trustworthy fashion. Or, all else equal, is it better to be a bit "panglossian" in our assumptions, giving others the benefit of the doubt? In many respects, this is the dilemma faced, of course, by the two research collaborators described at the beginning of this chapter.

3   Studying Trust Rules: "Thinking Aloud" about How People Think and Act in the Shadow of Doubt

There are many different ways a researcher can approach the task of trying to discover the cognitive rules people use when dealing with trust dilemmas. One approach, and in some respects a fairly direct method, is to simply ask people to tell us what they think about when they respond to various kinds of trust dilemmas. For example, to understand how negotiators assess others' trustworthiness in a negotiation context, we can ask them to "think aloud" about what they look for when sizing up a negotiator, and the behavioral rules they are likely to employ on the basis of that assessment.

One advantage of this direct approach is that it enables us to learn something about how people confronting a trust dilemma actually think. For example, we can learn something about the kinds of rules they believe will work well or poorly in the situation. We can also learn something about what they think other people are likely to do or not do in the situation (i.e., the common interpretation and action rules they expect others to use). This methodology also has the advantage of being inductive, thereby minimizing researcher assumptions about these important questions.

Using this thinking aloud approach, I examined how experienced organizational decision makers think about professional trust dilemmas. The study involved, in particular, interviews with 44 senior executives involved in an advanced management program (see Kramer, 2001 for a more complete description of the study rationale, methods and results). The executives were asked to talk about, among other things, how they would go about handling a trust dilemma involving a potential business partner whom they did not know much about (N.B. - the details of the vignette were structured so that they were quite similar to the dilemma described at the beginning of this chapter). They were asked a series of questions, for example, about the kinds of things they would look for when trying to assess the other's trustworthiness, and also the kinds of action rules (behaviors) they would use on the basis of that assessment.

One of the interesting patterns that emerged from this qualitative study was that it was possible to characterize about 80% of the respondents ($N = 35$) as having one of two general orientations towards other uncertainty about other people. The first group ($N = 23$), which I characterize as social optimists or panglossians, posess fairly positive or benign views about human nature. For these social panglossians, most people are fundamentally trustworthy most of the time. Although social uncertainty (uncertainty about their character, motivation, and intentions) is a part of social life, it is not something that has to be feared, but simply dealt with. Perhaps the biggest danger of such uncertainty, from the standpoint of these individuals, is that uncertainty introduces potentially disruptive "noise" into the communication process. As one executive in this category put it, "You have to be careful not to over-react to ambiguous information. Your assumptions about the other person can be wrong, and you can end up ruining a good relationship". Another individual put it this way, "You have to wait until all the facts are in. I give other people the benefit of the doubt until I'm reasonably sure. Getting a reputation for being a distrusting person would be fatal to my business, which runs largely on trust."

In contrast to these social panglossians, twelve individuals were categorized as having considerably more pessimistic and cautious views about other people and the effects of social uncertainty on their trust relations. As one put it, "I've always found it's better to play it safe than sorry. In my business, the costs of trusting the wrong person can be pretty severe. Because of the size of the deals I make, and the time horizon over which they play out, I need to make really good decisions up front". For these individuals, who I classify as social vigilants, social uncertainty is a problem. As one put it, "I always assume that people will try to take advantage of any loophole they find in a deal. You have to watch people when they think you aren't watching them – that's the time, in fact, you have to be most careful". According to these individuals, it is critical to pay attention to others, monitoring them closely for any evidence of lack of trustworthiness.

Interestingly, amongst both populations, there was considerable confidence expressed by executives as to the accuracy and adequacy of their views about human nature. As one executive put it, representatively, "I consider myself a pretty good judge of human nature". Another expressed the view, "I seldom make mistakes when sizing people up". Again, "I usually get a feeling pretty quickly about other people – about whether they can be trusted or not ... [that feeling] almost always turns out to be right". Viewed in aggregate, all of these quotes suggests how individuals who maintain fairly panglossian views about human trustworthiness versus those entertaining more pessimistic and paranoid views about such matters employ different interpretation and action rules when responding to social uncertainty. Yet, both groups are fairly confident their rules help them "navigate" successfully through the shoals of the trust dilemmas they confront in life.

Another approach to studying how our social auditor thinks and acts in the face of uncertainty–and in some respects, a more systematic approach–is to use computer simulation tournaments in which we "pit" different interpretation and action rule systems against each other. Using this approach, we can discover which cognitive rule systems are relatively robust and which tend to get people in trouble. In other words, we can learn something about which rule systems are good at generating mutually productive and trusting relationship and effective at minimizing exploitation, versus which rule systems tend to get people in trouble in terms of making mistakes about who to trust and how much. Thus, by allowing different rule systems to interact in a "noisy" (uncertain) social environment, we can learn something about the comparative efficacy of different rules and also the complex and often unintended interactions among them. The study I will briefly describe next, accordingly, took this approach.

Pragmatically, we can think about these matters from the standpoint of two decisions that the social auditor has to make in a trust dilemma situation. The first decision pertains to how much trust or distrust to display toward the other person at the very beginning of their relationship. The second pertains to how much adjustment to make in the trust or distrust displayed on the basis of feedback, however, uncertain or "noisy", regarding the other's actions.

To explore these questions from the perspective of decision makers possessing relatively panglossian versus paranoid rule systems, and to probe the consequences of these different auditing orientations, the study employed a computer simulation approach (see Bendor, Kramer, & Stout, 1991). The advantage of a computer simulation is that a researcher can precisely specify the nature of the dilemma (e.g., its "payoff structure"), and the consequences (costs and benefits) associated with any given choice. In other words, one can systematically evaluate how well different strategies for coping with trust dilemmas fare. Less obviously, a computer simulation also provides an indirect way for trust researchers to study people's beliefs about trust dilemmas because, in designing a rule system for "playing" in a computer simulation tournament, designers have to decide how others will respond to social uncertainty or "noise" (e.g., will they try to exploit it by "sneaking around" under it or work to keep misunderstandings from developing?).

The computer simulation we used was similar to that employed by Axelrod (1984) in his pioneering studies on the evolution of cooperation. Recall that, using a tournament approach, Axelrod had individuals design strategies that would play other strategies in a classic iterated and deterministic (noiseless) Prisoner's Dilemma Game. Because we were interested in studying how social uncertainty affects decisions, however, the procedures we used differed from Axelrod's procedures in several important ways. First and foremost, we introduced uncertainty or "noise" into the communication process between the interdependent social actors. In the classic Prisoner's Dilemma, after interactants decide whether they wish to cooperate with the other, they learn with perfect clarity whether their partners cooperated or not. As noted earlier, in trust dilemma situations, people suffer from some degree of uncertainty regarding what their partners did.

Accordingly, in order to introduce social uncertainty into our tournament, participants' effort levels were obscured by adding or subtracting a small random amount of "noise" to their feedback during each period of play. This noise term was a random variable, distributed normally, with a mean of zero and a standard deviation of eight. The term was generated by a normal transformation of a uniform random variable and the distribution of the error term was truncated to the interval $[-100, +100]$. The noise was independent across participants and over periods. Because of this noise factor, participants in our tournament (or more precisely, the computer programs of the strategies they submitted) could receive feedback that their partners (strategies) had behaved either more or less cooperatively than they actually had. This allowed us to explore how decision makers "adjusted" to the noise (e.g., did they adopt relatively panglossian or paranoid views about the noise?).

Adjustments to the noise were manifested in terms of how much aid or help decision makers wanted to extend to their partners. This was operationalized as how much effort they wanted to exert on their partner's behalf, with zero effort being the least effort extended and 100 percent effort being the highest possible level of effort. In this way, we tried to capture the idea that people typically make decisions about how much trust or distrust is warranted, especially in a new relationship, where one possesses little prior experience with the partner.

A player's payoff or benefit per interaction, therefore, was equal to the other player's actual effort level minus a cost factor of one's own effort plus a random error term. Stated more formally,

$$V_t(i/j) = C_t(j,i) - \alpha C_t(i,j) + \varepsilon_t(j,i) \tag{1}$$

where $V_t(i/j)$ denotes i's payoff in period t in its pairing with j, $C_t(j,i)$ denotes j's effort level vis-a-vis i in period t, $C_t(i,j)$ refers to i's choice, α the cost of helping associated with i's choice and $\varepsilon_t(j,i)$ symbolizes the disturbance added to j's choice. To ensure the game meets the requirements of a Prisoner's Dilemma, the cost-of-helping parameter must be between zero and one; accordingly, it was fixed at 0.80. With this structure, the symmetric average maximum payoff per period was 20 and the symmetric average minimum payoff was 0. However, with noise, realized payoffs could fall outside this region of feasible expected payoffs.

In each round of play, the computer generated two normally distributed error terms to determine the values of $\varepsilon_t(i,j)$ and $\varepsilon_t(j,i)$. The computer also used a fixed random number which was used to determine whether a given sample path would end after the current period of play (so that the duration of a relationship was also uncertain). Each strategy ten played all other strategies, including itself, using the generated error terms for each period.

Entrants to the tournament were fully informed of all of the parameters described above, including details about payoffs, the distribution of the random disturbances, and the stopping probability of the game. However, they were also instructed that they (i.e., their strategies) would not be told the realized values of the disturbances, their partner's true effort level, or the rules defining their partner's strategy. They were then invited to submit strategies which would be translated into computer programs. The computer programs would be pitted against each other in pairwise play using a round robin tournament format.

From the standpoint of thinking about coping with social uncertainty, we can think of the social auditor's cognitive rule systems as having to confront two decisions. The first decision is how much presumptive trust in the other to manifest initially. For example, should one extend full effort to the other initially in the hope that the other will reciprocate with full effort? Extending full aid initially will help sustain a mutually productive relationship with a similarly inclined other. On the other hand, it opens one up to early exploitation if one happens to encounter a more predatory other. Thus, given the inherent uncertainty in the situation, it shows a high level of presumptive trust in the other. Alternatively, should one display some level of scepticism or wariness regarding others willingness to reciprocate our trust in them, withholding aid to them until we see what they are willing to do? Caution provides protection against initial exploitation, but its risks damaging a relationship with some one who be equally willing to extend full effort to us.

The second decision is how to respond to feedback regarding the others' actions–especially when one knows that such feedback is inherently uncertain (contaminated by noise). Stated differently, how much sensitivity should one display to indications that the other might be reciprocating fully or less than fully given those indications are inherently ambiguous? How reactive should one be, for example, to the hint or suspicion that the other might not be reciprocating fully?

To explore these issues, we sought to recruit participants who were sophisticated decision makers (i.e., individuals would be intimately familiar with theory and research on the Prisoner's Dilemma). Thus, the participants included leading game theorists, economists, political scientists, and social psychologists. Note that we can view the strategies designed by these experts as, in a sense, their cognitive rule systems that reflect their "projections" about others' orientation towards such dilemmas. In other words, they represent their orientations, as social auditors, to the dilemma in question.

Thirteen strategies were submitted to the tournament, embodying a variety of different intuitions regarding how decision maker should cope with social uncertainty. For the purposes of the present chapter, I will compare and contrast the performance of a strategy I call VIGILANT with that of a strategy I call

PANGLOSSIAN. VIGILANT, to be described first, is our relatively paranoid social auditor.

In terms of the decision rules embodied in this strategy, VIGILANT was nice in the Axelrodian sense of that term (i.e., it always started out by extending maximum effort to its partner). It was therefore willing to initially assume the possibility that it was dealing with another strategy willing to reciprocate fully. In the parlance of the intuitive auditor model, it was willing to show some presumptive trust toward the other. However, as its name suggests, it was extremely attentive and reactive to any signs that its partners were not reciprocating fully. In particular, if VIGILANT detected what it believed was less than full effort from any of its partners on any trial, it retaliated by completely withdrawing effort (i.e., giving 0 points to its partner) on the next trial. The aim of this extreme reactivity was, of course, to deter further acts of perceived exploitation. Stated in psychological terms, VIGILANT was determined to minimize the costs associated with misplaced trust.

Because VIGILANT was determined not to be exploited, it was biased, in one sense, toward "over assuming" the worst whenever it received a low offer of effort from its partner. In other words, whenever it received feedback that the other had not offered much, it acted as if its partner had intentionally trying to short-change it. Presuming this interpretation to be valid, VIGILANT then retaliated. Thus, VIGILANT always puts the least charitable "spin" on any social news it receives about others' actions. It thus goes through life assuming the worst with respect to others, even though there is a clear and compelling rival explanation for such outcomes (i.e., it's a noisy world and sometimes one will simply have bad luck when communicating with others).

There are different ways of dissecting VIGILANT's performance. If we start with Table 1, we can see how VIGILANT fared against the other players. In particular, if we first compare the row payoffs in Table 1, which reflect how much help VIGILANT received on average from the other players over the course of its interactions, to the column payoffs, which reflect how much help VIGILANT it offered to these players, we observe a powerful pattern, viz., that VIGILANT always got back more than it gave away. In terms of benefits accrued versus those given, VIGILANT walked away a clear "winner" in every one of its close encounters. It outperformed every strategy in terms of comparative payoffs. In other words, against an array of clever players, it came out ahead of the game in every instance. VIGILANT seems to bat a 1000.

If we think about VIGILANT'S performance from the standpoint of our social auditor model, we would have to conclude that here is a decision maker who is likely to be rather satisfied with his performance in this world. He has, after all, come upon what seems to be a very satisfactory, even powerful set of cognitive rule. Being vigilant appears to payoff handsomely; it consistently elicits reliably good against the full panoply of players, and avoids the prospect of misplaced trust. No one it meets does better than it does: it walks away from every encounter taking out from more than it put into the relationship. It is hard to imagine such a decision maker experiencing much cognitive dissonance

Table 1.

	NICE	DRIFT.	BTFT	MENN.	W.A.	TF2T	Even	TFT	Norm.	Run.	Dev.	CTTF	VIG.
PANGLOSSIAN	19.8	12.0	19.5	19.5	18.5	18.2	19.3	17.4	18.4	14.0	17.1	13.0	14.9
DRIFTING	26.0	16.8	23.7	23.7	18.1	20.5	17.6	17.5	19.2	12.3	12.4	6.7	0.7
BTFT	20.0	13.5	19.7	19.7	18.7	18.5	19.4	17.5	18.5	13.0	15.9	12.6	3.1
MENNONITE	20.0	13.5	19.7	19.7	18.7	18.5	19.4	17.5	18.5	12.1	15.7	12.6	3.8
Wtd. Avg.	20.8	15.7	20.4	20.4	17.9	19.3	18.4	17.0	18.7	12.5	18.0	7.3	1.7
TF2T	21.0	15.1	20.5	20.5	18.7	18.9	19.0	17.1	18.3	10.4	13.7	10.6	-0.2
Staying Even	20.2	16.2	19.9	19.9	18.3	19.0	19.3	15.4	18.0	13.4	12.1	6.2	2.2
NAIVE REALIST	21.7	16.2	21.1	21.1	17.7	19.1	15.9	14.2	15.9	13.1	11.5	6.6	1.0
Normal	20.9	15.2	20.3	20.3	18.3	18.6	18.1	14.8	16.4	11.3	11.9	7.4	-1.4
Running Avg	17.0	11.5	18.0	19.1	14.8	18.3	14.3	14.4	17.0	12.8	6.2	10.4	9.0
Deviations	21.9	11.6	20.0	20.4	17.8	16.5	12.8	12.4	15.0	7.4	14.4	7.2	4.7
CHEATING TFT	25.1	7.8	22.7	22.7	12.3	17.4	8.7	9.4	11.4	13.7	10.9	5.8	-0.4
VIGILANT	23.6	3.2	9.9	9.0	9.1	9.8	5.2	6.5	10.7	13.9	8.3	10.7	2.6

– the rule system it is employing seems, in short to be working. It is possible to imagine that VIGILANT can even feel rather virtuous about its values, given its view of the world. It is, after all, nice (i.e., always willing to initially extend full aid to everyone it meets). It's also strong and affirmative (i.e., unwilling to tolerate exploitation or abuse). Yet, although tough, it is also forgiving: it is willing to let bygones be bygones.

Let's turn attention now to the performance of our second strategy, called PANGLOSSIAN, which embodies the more benign view of the world associated with the social optimists described earlier. In terms of its rule structure, PAN-GLOSSIAN, like VIGILANT, was nice in that it always began a relationship by extending maximum possible aid to its partner. However, it differed from VIGILANT in several ways. First, PANGLOSSIAN tended to be generous, re-turning more effort on a subsequent trial than it had received from its partner on the previous trial. PANGLOSSIAN's generosity took the form of what might be construed as a form of benign indifference (as long as it partner's observed (realized) effort level exceeded 80, PANGLOSSIAN would continue to help its partner fully, i.e., give 100% of the possible points to it. Second, although PAN-GLOSSIAN was provocable and would retaliate if it's partner's aid dipped below 80, it reverted to extending full effort to its partner again, as long as the partner satisfied this threshold of acceptable behavior (i.e., its observed aid level was at least 80). Thus, it could be provoked, but was forgiving. It was thus less reactive than VIGILANT (slower to anger) and quicker to forgive.

If we compare how well PANGLOSSIAN did against the other players, we observe another seemingly clear and compelling pattern. As is quite evident from inspection of the results shown in Table 1, PANGLOSSIAN lost out in its encounters with every player in the tournament. In other words, it always gave away to its partners more than it got back from them. In terms of its performance, then, PANGLOSSIAN seems to be a very poor judge of others and clearly sub-optimally adapted to the world in which it finds itself (especially compared to VIGILANT). It appears to be far too generous, consistently over-benefitting its partners.

In terms of the logic and structure of the social auditor model, note that PANGLOSSIAN tends to give others the benefit of the doubt when it comes to drawing inferences about what low extended effort from the other really means. In particular, it attributes low returns to the environmental factor (bad luck with respect to the noise) rather than necessarily assuming it was the intention of the other player, as does VIGILANT. Thus, as long as it is getting at least 80% of the possible payoffs, it tends to under-react to unfavorable "news" from its environment. This under-reaction seems to be rather costly, however, allowing it to be bested not only by more suspicious or wary players such as VIGILANT, but in fact every player in the tournament.

At first glance, therefore, the comparison of these two social auditing strategies and rule systems seems to suggest a rather compelling conclusion: in an uncertain world, a certain amount of suspicion and wariness – coupled with a willingness to react strongly to even ambiguous evidence of others' untrustworthiness – can be deemed prudent and even beneficial. Certainly, VIGILANT appears to be less likely to be taken advantage of, whereas PANGLOSSIAN seems to turn the other cheek too often, far and for too long.

If we inspect the tournament results from a different vantage point, however, the story told by these tournament data turns out to be less simple and, in fact, quite a bit more interesting. In particular, if we compare the overall performance of the strategies (i.e., examine their total earnings in the tournament by summing their average payoffs across encounters with all strategies), we see a dramatically different picture (see Table 2).

Table 2.

Rank	Name	Average per Period Payoff
1	PANGLOSSIAN	17.05
2	DRIFTING	16.55
3	BIASED TFT	16.17
4	MENNONITE	16.13
5	WTD. AVE RECIP	16.00
6	T2FT	15.67
7	STAYING EVEN	15.39
8	NAÏVE REALIST	15.00
9	NORMAL	14.78
10	RUNNING AVERAGE	14.06
11	DEVIATION WITH ANCHOR	13.74
12	CHEATING TFT	12.88
13	VILIGANT	8.99

As Table 2 shows, PANGLOSSIAN emerges as the tournament winner when we use this standard of performance. PANGLOSSIAN was the most successful performer, earning an average per trial payoff of 17.05 points. Thus, the performance of PANGLOSSIAN outshines that of VIGILANT when looked at from this perspective. Yet, recall, VIGILANT outperformed PANGLOSSIAN in their "face-to-face" encounter with each other, and also every other player as well, whereas PANGLOSSIAN lost out against every player!

How can VIGILANT do better in every encounter with every player it meets in its world, and nonetheless end up losing the tournament, even coming in "dead last", while PANGLOSSIAN gets the "short end of the stick" in every encounter in the same world and yet ends up accumulating more resources than any other player in the tournament? This pattern of results – at least at first glance – seems paradoxical. The answer to the riddle can be found, however, by re-visiting Table 1 and inspecting the absolute pay offs each strategy obtains from the other strategies, rather than just looking at its comparative (relative) payoffs. In particular, if we compare the absolute means of PANGLOSSIAN's payoffs from its encounter with each player to those garnered by VIGILANT we see that PANGLOSSIAN walks away from each relationship with rather respectable payoffs. Indeed, many times it manages to elicit almost the maximum possible benefit even allowing for the presence of noise (approaching close to the maximal average mean of 20).

VIGILANT, in contrast, earns much less in absolute terms from these same encounters. It may walk away the winner from each encounter in its social world when we think only in social comparative terms (i.e., how much it did relative to its partner), but it lives a life of impoverished relationships and diminished returns when we look at the absolute yield it gets in those relationships. In contrast, PANGLOSSIAN may walk away from each relationship a bit worse off in terms of the "giving-getting" equation, but it manages to extract a lot of absolute net gain from each encounter. Thus, in a powerful way, these findings illustrate how different cognitive rule systems, embodying different auditing orientations and principles, can produce very different experiences in life. One auditing orientation does well from the perspective of a myopic and narrow social accounting scheme that highlights proximate payoffs defined at the level of local social comparisons. The other does well from the perspective of a long-term perspective that values absolute (noncomparative) payouts from relationships.

As an additional note, it is instructive to examine how well the strategy labeled in Tables 1 and 2 as a strict reciprocator called NAÏVE REALIST fared in this tournament. NAÏVE REALIST adjusted to the noise factor by simply returning to its partner whatever it received from them on the previous trial. Since the average value of the noise term was 0, it assumed on average this was a reasonable stance to take toward social uncertainty. NAÏVE REALIST is equivalent to Axelrod's TIT-FOR-TAT). Recall that Axelrod's results had demonstrated the power of strict reciprocators such as TIT-FOR-TAT in a deterministic game, where decision makers have perfect (noiseless) information about a partner's actions. Yet, in a tournament climate clouded by uncertainty, we found that this strategy suffered a sharp decline—both absolutely and relative to other players. Indeed, NAÏVE REALIST placed a distant eighth in the final rankings, earning only 75% of the maximum symmetric payout. Thus, in a world of uncertainty, strict auditing and reciprocity apparently get one in trouble.

What accounts for this degraded performance of the strict reciprocator in a world filled with social uncertainty? Why did a strategy like PANGLOSSIAN, with its more benign assumptions, do better than the more realistic auditors?

Part of the answer to such questions can be gleaned by observing how a strict reciprocator such as NAÏVE REALIST behaves when it plays itself in the presence of uncertainty. Because NAÏVE REALIST is nice, it starts off by extend full effort to its partner. Sooner or later, however, an unlucky ("bad") realization of the random error term occurs. Because NAÏVE REALIST is provocable, it will retaliate in the next period. Its partner, being similarly provocable, returns the compliment, leading to cycles of counterproductive mutual punishment, resulting in steadily falling levels of effort. In contrast, more generous strategies such as PANGLOSSIAN slow down this degradation by returning more than an unbiased estimate of their partner's effort level. This generosity tends to dampen these cycles of unintended and costly vendettas.

The surprising performance of PANGLOSSIAN, especially when contrasted with what might seem, at first glance to be fairly vigilant and tough auditors such as NAÏVE REALIST and VIGILANT, merits more sustained analysis. Why does PANGLOSSIAN's generosity tend to work better than strict reciprocity in a noisy setting? Stated differently, why do rules systems that favor vigilance and strict reciprocity fare so poorly in an uncertain world? PANGLOSSIAN benefits by its not being envious of what its partner gets. As Axelrod (1984) had noted, envy can get decision makers in trouble if it sets off cycles of mutual discontent and retaliation. By setting an absolute (nonsocial comparative) standard for what it considers a reasonable rate of return on its relationships, PANGLOSSIAN does not pay any attention to what its partner gets. In contrast, VIGILANT displays a great deal of concern about such comparisons – at least, it does not wish to grant full aid to others when it suspects it is not getting full aid back from them in return. Over the long haul, it pays a steep price for its insistence on never getting the "short end of the stick" in its relationships.

Note also that, by under-reacting to the noise, PANGLOSSIAN accomplishes something else – something that is important but easy to overlook. It solves the other actor's trust dilemma. In other words, by keeping its aid levels high, its partner does not have to decide whether or not to punish PANGLOSSIAN. Recall that receiving any aid from the other above 80% is coded as justifying or warranting returning full aid back to the other. This point is important because in much of the trust dilemma literature, the focus is often on how social actors can solve their own trust dilemma. Little attention has been given to the importance of solving the other person's trust dilemma. Yet, by shifting focus in this way, we may buy ourselves something more powerful. In a sense, we make it easier for them to be good to us by making trusting behavior easier to justify. We reduce their perception of vulnerability and end up reducing our own as well. This point is especially important because of the insidious effects of self-serving and self-enhancing judgmental biases in such dilemmas (Kramer, Newton, & Pommerenke, 1993).

4   How the Intuitive Social Auditor Gets into Trouble: People's "Naïve Theories" about the Efficacy of Different Orientations and Rule Systems

As noted earlier in this chapter, most people articulate rules they use in trust dilemma situations quite readily. For example, when executives are asked to think about these situations in real-organizations, they are able to enumerate a number of general "rules of thumb" governing the building of trust and/or protecting themselves against the prospect of misplaced trust. Moreover, they apparently have firm ideas about how others think about social conduct and how they are likely to behave as well. In short, people seem to easily assume the role of both philosopher and psychologist: they ponder about both what is right (the social decision heuristics one ought to use in a given situation) as well as what is effective (the heuristic that is likely to produce the "best" outcome in that situation). But how well do such beliefs and intuitions correspond to the empirical conclusions suggested by our noisy tournament? Recall that the computer simulation study just described, we had recruited leading game theorists, psychologists, economists, and political scientists. These were very smart people and people able to think deeply and with nuance about the strategic and complex interactive properties of these situations. Yet, even with this elite sample, often their strategies did not perform well against even their simple clone. Thus, even a fairly simple first-order mental simulation of strategic interaction might have been difficult (although it is possible that they recognized that, even though their strategies might not well against each other, they would do well against the other strategies submitted in the tournament!).

Accordingly, to explore decision makers' intuitions or "naïve theories" about the fairness and efficacy of different heuristics, we investigated MBA students' beliefs about PANGLOSSIAN, VIGILANT, and NAÏVE REALIST. We gave students enrolled in an elective course complete information about all of the parameters for our noisy computer tournament. In fact, their information was identical to the information given to the entrants in the original computer simulation study. We then asked them to predict how well the three strategies (NAÏVE REALIST, PANGLOSSIAN, and VIGILANT) would do, i.e., to predict the average payoff of each entry. We also asked students in a second class to predict the relative performance or ranking of the three strategies.

As can be seen from Table 3, the results suggest that students' intuitions about the comparative efficacy of these three strategies closely parallel the findings of Axelrod's original tournament. They clearly expected TFT to be the most efficacious in terms of their expectations about both absolute and relative performance. VIGILANT was also viewed as a fairly effective strategy. Significantly, PANGLOSSIAN ran a distant third in student's expectations and predictions.

To probe further people's intuitions about these rule systems – and how well they help or hinder one's success in life – I asked participants in an executive program to write short paragraphs describing their beliefs about how strategies would behave and perform. One participant noted about PANGLOSSIAN that,

Table 3. Predicted Performance of Heuristics*

Heuristic	Predicted Mean Score**	Predicted Rank***		
		1	**2**	**3**
NAIVE REALIST	18.35	42	17	7
VIGILANT	17.64	18	44	4
PANGLOSSIAN	15.66	6	5	55

* Predicted scores and ranks provided by separate samples
** N = 63
*** N = 66

"If you're generous, you probably will get screwed [out of your payoffs] too much. You'd be out of business in no time". In commenting on the attractiveness of NAÏVE REALIST, in contrast, another wrote, "The nice thing about NAÏVE REALIST is that it keeps the playing field level – and that's important over the long term". "By far, NAÏVE REALIST is the fairest strategy". And in thinking about the advantages of VIGILANT, a participant stated, "In many situations . . . you've got to cover your back side". As former CEO of Intel, Andrew Grove likes to remind his employees, "Only the paranoid survive".

5   Implications and Conclusions

Viewed together, the results from our examination of people's lay theories and the results of the computer simulation indicate there is sometimes a disparity between people's beliefs about the efficacy of different auditing orientations and rule systems, and their actual performance (at least in some ecologies). One reason why people's beliefs may be at least partly wrong may be that they worry more about (and therefore overweight) the possibility of being exploited, while under appreciating the mutual gains produced by generosity. Comparing the long-term performances of PANGLOSSIAN and VIGILANT illustrates how complex and sometimes counter-intuitive are the tradeoffs between short-term and long-term results.

An important question for future research to engage is, "When are auditing and rule systems predicated on generosity more psychologically appealing and sustainable than those predicated on paranoia and "strict" reciprocity?" It seems reasonable to argue, on prima facie grounds, that a number of factors contribute to the attractiveness of generosity and lessen the perceived need for the strict auditing and balancing-of-accounts required by VIGILANT and NAÏVE REALIST. First, a shared identity among social actors may help. Decision makers are probably more willing to behave generously when interacting with others with whom they share a social tie or bond. For example, when decision makers from the same neighborhood meet face to face, they may trust each other more and cooperate more reliably than they would otherwise (cf., Ostrom, 1998). Further, common identification with an ongong partner along some salient dimension, such as a common group identity, may affect which strategy is viewed as most fair (Kramer, Pommerenke, & Newton 1993) and/or most effective (Rothbart & Hallmark, 1988). Research on moral exclusion (Opotow, 1990) suggests that individuals impute higher standards of fairness to those whose group identity they share. Consistent with Opotow's proposition, Kramer et al. (1993) found that increasing the salience of shared identity enhanced concerns about equality of outcomes.

Taking the other party's perspective may also influence individuals' willingness to be generous in dilemma-like situations. Arriaga and Rusbult (1998) have found that the experience of "standing in a partner's shoes" increases constructive responses to the accomodative dilemmas they jointly experience, at least in close personal relationships. In line with this hypothesis, it is interesting to note that students of the Cuban Missile Crisis have pointed out the restraint President Kennedy showed during that Crisis. When choosing to under-react to some Soviet action that seemed to call for a Tit-for-Tat-like response, President Kennedy tried to put himself in Khruschev's shoes. He sought to understand how the situation in Cuba might look to Khruschev, to grasp the kinds of political pressures and institutional momentum driving his decisions (Garthoff, 1989).

Due to the actor-observer bias, individuals are likely to construe their partner's noisy realizations in dispositional terms, while underestimating the impact of the noise as a situational factor. In contrast, they are likely to assume that their partner properly understands the situational constraints and liabilities under which they labor (cf., Jervis, 1976).

In putting these results in perspective, it is important to note that it is easy to overweight the adaptive value of panglossian world views, just as it would be easy to underweight the functional advantages of paranoid views (cf., Kramer, 1998). The virtues of generosity emerge only in a world in which there are enough other strategies to make generosity pay (and stinginess hurt). In nastier ecologies, where the costs of misplaced trust are steep, being very generous may put one on the path to peril. In nastier ecologies, never walking away the loser may be important for establishing reputational deterrence – leading to longer cumulative gains than nicer, more generous strategies that invite exploitation and attract predators. Thus, the wisdom of turning the other cheek depends ultimately on the distribution of dispositions in one's social world. Accepting the fundamentally contingent nature of adaptive auditing and action in trust dilemma situations may be psychologically unpleasant but necessary.

References

1. Allison, S. & Messick, D., (1990), *Social decision heuristics in the use of share resources*, Journal of Behavioral Decision Making, 3, 195-204, 1990.

2. Arriaga, X. B., & Rusbult, (1998), C. E., *Standing in my partner's shoes: Partner perspective taking and reactions to accommodative dilemmas*, Personality and Social Psychology Bulletin, 24, 927-948.
3. Axelrod, R., (1980a), *Effective choice in the prisoner's dilemma*, Journal of Conflict Resolution, 24, 3-25.
4. Axelrod, R. (1980b), *More effective choice in the prisoner's dilemma*, Journal of Conflict Resolution, 24, 379-403.
5. Axelrod, R. (1984), *The evolution of cooperation*, New York: Basic Books.
6. Axelrod, R. & Dion, D. (1988), *The further evolution of cooperation*, Science, 242. 1385-1390.
7. Baron, J. (1996), *Do no harm*, In D. M. Messick & A. E. Tenburnsel (Eds.), Codes of conduct: Behavioral research into business ethics, pp. 197-214. New York: Russell Sage Foundation.
8. Bazerman, M. (1986), *Judgment in managerial decision making*. New York: Wiley.
9. Bazerman, M. H. (1994), *Judgment in managerial decision making*. New York: John Wiley.
10. Bendor, J. (1987), *In good times and bad: Reciprocity in an uncertain world*, American Journal of Political Science, 31, 531-558.
11. Bendor, J., Kramer, R. M., & Stout, S. (1991), *When in doubt: Cooperation in the noisy prisoner's dilemma*. Journal of Conflict Resolution, 35, 691-719.
12. Bendor, J., Kramer, R. M., & Swistak, P. (1996), *Cooperation under uncertainty* (Comment on Kollock), American Sociological Review, 61, 333-338.
13. Cialdini, R. (1993), *Influence: Science and Practice*, New York: Harper-Collins.
14. Garthoff, R. (1989), *Reflections on the Cuban Missle Crisis*, Washington, D.C.: The Brookings Institution.
15. Gigerenzer, G., & Too, P. M. (1999), *Simple heuristics that make us smart*, Oxford, England: Oxford University Press.
16. Jervis, R. (1976), *Perception and misperception in international politics*, Princeton, NJ: Princeton University Press.
17. Kramer, R. M. (1996), *Divergent realities and convergent disappointments in the hierarchic relation: The intuitive auditor at work*, In R. M. Kramer and T. R. Tyler (Eds.), Trust in organizations: Frontiers of Theory and Research. Thousand Oaks: Sage Publications.
18. Kramer, R. M. (1998), *Paranoid cognition in social systems*, Personality and Social Psychology Review, 2, 251-275.
19. Kramer, R. M. (2001), *Trust and distrust as adaptive orientations in trust dilemmas*, Unpublished manuscript.
20. Kramer, R. M. (1999), *Trust and distrust in organizations: Emerging perspectives, enduring questions*, Annual Review of Psychology, 50, 569-598.
21. Kramer, R., Meyerson, D. & Davis, G. (1990), *How much is enough? Psychological components of 'guns versus butter' decisions in a security dilemma*, Journal of Personality and Social Psychology, 58, 984-993.
22. Kramer, R., Newton, E., & Pommerenke, P. (1993), *Self-enhancement biases and negotiator judgment: Effects of self-esteem and mood*, Organizational Behavior and Human Decision Processes, 56, 110-133.
23. Kramer, R., Pommerenke, P., & Newton, E. (1993), *The social context of negotiation: Effects of social identity and accountability on negotiator judgment and decision making*, Journal of Conflict Resolution, 37, 633-654.
24. March, J. G., Schulz, M, & Zhou, X. (2000), *The dynamics of rules: Change in written organizational codes*, Stanford, CA: Stanford University Press.

25. Molander, P. (1985), *The optimal level of generosity in a selfish, uncertain environment*, Journal of Conflict Resolution, 29, 611-618.

26. Novak, M. & Sigmund, K. (1992), *Tit-for-Tat in heterogeneous populations Nature*, 355, 250-253.

27. Novak, M. & Sigmund, K. (1993), *A strategy of win-stay, lose-shift that outperforms tit-for-tat in the prisoner's dilemma game*, Nature, 364, 56-58.

28. Opotow, S. (1990), *Deterring moral exclusion*, Journal of Social Issues. 46, 173-182.

29. Ostrom, E. (1998), *A behavioral approach to the rational choice theory of collective action*, Presidential address, American Political Science Association, 1997. American Political Science Review, 92, 1-22.

30. Ross, L. (1977), *The intuitive psychologist and his shortcomings*, In L. Berkowitz (Ed.). Advances in experimental social psychology, 10. New York: Academic Press.

31. Thompson, L. (1998), *The mind and heart of the negotiator*, Upper Saddle River, NJ: Prentice Hall.

32. Watson, J. (1980), *The double helix: A personal account of the discovery of the structure of DNA*, (Norton Critical Edition). New York: W. W. Norton.

33. Wilson, J. Q. (1993), *The moral sense*, New York: Free Press.

34. Wu, J., & Axelrod, R. (1997), *Coping with noise: How to cope with noise in the iterated prisoner's dilemma*, In R. Axelrod (Ed.), The complexities of cooperation.

Trust and Distrust Definitions:
One Bite at a Time

D. Harrison McKnight[1] and Norman L. Chervany[2]

[1] Accounting and Information Systems Department
The Eli Broad Graduate School of Management
Michigan State University, East Lansing
MI, 48824-1121, USA
mcknight@bus.msu.edu
[2] Information and Decision Sciences Department
Carlson School of Management
University of Minnesota
Minneapolis, Minnesota, 55455
nchervany@csom.umn.edu

Abstract. Researchers have remarked and recoiled at the literature confusion regarding the meanings of trust and distrust. The problem involves both the proliferation of narrow intra-disciplinary research definitions of trust and the multiple meanings the word trust possesses in everyday use. To enable trust researchers to more easily compare empirical results, we define a cohesive set of conceptual and measurable constructs that captures the essence of trust and distrust definitions across several disciplines. This chapter defines disposition to trust (and -distrust) constructs from psychology and economics, institution-based trust (and -distrust) constructs from sociology, and trusting/distrusting beliefs, trusting/distrusting intentions, and trust/distrust-related behavior constructs from social psychology and other disciplines. Distrust concepts are defined as separate and opposite from trust concepts. We conclude by discussing the importance of viewing trust and distrust as separate, simultaneously operating concepts.

" . . . *trust is a term with many meanings.*" – *Oliver Williamson*
"*Trust is itself a term for a clustering of meanings.*" – *Harrison White*
"*. . . researchers . . . purposes may be better served . . . if they focus on specific components of trust rather than the generalized case.*" – *Robert Kaplan*

1 Introduction

The human drama often involves parties who trust and distrust each other at the same time. For example, during World War II, Franklin D. Roosevelt and Joseph Stalin had to trust or rely on each other for mutual support and cooperation against a common foe, while at the same time distrusting each other's actions because each knew that the other had his own interests to serve. Even

R. Falcone, M. Singh, and Y.-H. Tan (Eds.): Trust in Cyber-societies, LNAI 2246, pp. 27–54, 2001.
© Springer-Verlag Berlin Heidelberg 2001

though they were guided by different ideologies, each believed the other would display enough integrity to fulfil agreements they made to conduct the war in certain agreed-upon ways. They were therefore willing to depend on each other and actually depended on each other (i.e., they trusted), even though they were aware of potential problems in their relationship. Trust and distrust are widely acknowledged to be important or even vital in cooperative efforts in all aspects of life [1, 13, 15], including organizations [22]. If trust and distrust are important, some effort should be devoted to defining them. We justify and specify a conceptual typology of high level trust and distrust concepts. Then we define, as subsets of the high level concepts, measurable constructs for empirical researchers. These definitions are not meant as a prescribed set to use in a given study, but are intended as a menu of clearly defined trust concepts from which researchers may select. Although we define trust in terms of people, these definitions may be adapted for trust of people in computers or trust between computer agents.

2   The Challenge of Conceptualizing Trust and Distrust

In spite of trust's import, trust research efforts are sometimes hard to follow and difficult to compare with each other because the term trust is defined in a multitude of different ways [35, 93], such that researchers have marveled at how confusing the term has become [3, 86]. Trust has not only been described as an "elusive" concept [103: 130], but the state of trust definitions has been called a "conceptual confusion" [50: 975], a "confusing potpourri" [86: 625], and even a "conceptual morass" [3: 1, 10: 473]. For example, trust has been defined as both a noun and a verb (e.g., [3]), as both a personality trait [78] and a belief [52], and as both a social structure [86] and a behavioral intention [12, 84]. Some researchers, silently affirming the difficulty of defining trust, have declined to define trust, relying on the reader to ascribe meaning to the term (e.g., [32, 71]).

Whereas it is arguably more important to conceptual clarity to understand the essence of what trust is than how it forms, some researchers have primarily defined trust types in terms of the bases by which trust forms. Zucker's [105] typology included process-based, characteristics-based, and institutional-based trust. Shapiro, Sheppard and Cheraskin [88] and Lewicki and Bunker [48] espoused calculus/deterrence-based, knowledge-based, and identification-based trust. However, because these typologies focus on trust's bases rather than on what trust means, they do not address the conceptual clarity of trust directly. Still, they improve our understanding of trust, just as did definitions of the bases of power (e.g., [23]) in the early power literature.

Although the term distrust has been researched less than has trust, it too needs to be conceptually clarified. Distrust (sometimes called 'mistrust') has been defined to have widely different meanings. For example, distrust means: a "belief that a person's values or motives will lead them to approach all situations in an unacceptable manner" [90: 373], as an expectation "of punishments from Other ... rather than rewards" [84: 77], or as a choice to avoid a risky, ambiguous path [15]. We will argue later that trust and distrust are separate constructs that may exist simultaneously. Distrust is not only important because it allows one to avoid negative consequences, but because general distrust of other people and institutions is becoming more prevalent [15, 67, 78], which means that it may, to an extent, be displacing trust as a social mechanism for dealing with risk. Indeed, under certain conditions, distrust may already be more useful or beneficial than trust. However, without properly defining trust and distrust, it would be hard to tell which is more important-and when.

To make progress in trust/distrust research requires a hard look at what trust and distrust mean, for Niklas Luhmann said that in order to gain greater insights about the nature of trusting relations, "we need further conceptual clarification" [55: 94]. Because effective conceptualization is a necessary condition for producing good research [40], creating good conceptual definitions of trust and distrust should take priority over both theory-testing studies and psychometric measurement studies of trust [40, 85]. Wrightsman [102], a trust scholar himself, pointed out that researchers need a good model of trust constructs because measurement of trust has clearly outstripped adequate conceptualization.

Like an elephant, the large, unwieldy topic of trust has been hard for researchers to get their arms around for at least two reasons. First, much of the confusion about trust has resulted from the divergent *weltanschauungs* (world views) of various intellectual disciplines. Like the story of the six blind men, who each described the elephant based on the portion of the elephant's body they touched, each research discipline has applied its own lens to one part of the trust elephant's anatomy [48]. A disciplinary lens sheds significant light on a topic like trust, but can also blind the researcher to possibilities outside the paradigm the discipline pursues [4]. Based on the differences among their definitions of trust, it appears that psychologists analyzed the personality side, sociologists interviewed the social structural side, and economists calculated the rational choice side of the trust elephant. Few researchers, such as [3, 9, 24, 41, and 66], have developed trust typologies that define a set of trust constructs, and fewer still, such as [57, 62], have tried to reconcile interdisciplinary sets of constructs. More typically, trust typologies have stubbornly retained an intra-disciplinary flavor (e.g., [41, 51]. As Doney, Cannon, and Mullen stated, "Developing an integrated model of trust is particularly difficult, given the vagueness and idiosyncrasies in defining trust across multiple disciplines and orientations" [16: 603].

Second, trust is conceptually, like the elephant, massive in terms of the meanings it conveys. In everyday usage, trust has more dictionary definitions than do the similar terms 'cooperation', 'confidence', and 'predictable' combined. Mayer, et al. [57] used these terms to discriminate trust from similar concepts. An analysis of the word trust in three unabridged dictionaries (Websters, Random House, and Oxford) showed that trust had far more definitions (9, 24, and 18, respectively) than did the terms cooperation (3, 2, 6), confidence (6, 8, 13), and predictable (1, 2, 1). On average, trust had 17.0 definitions, while the others had an average of 4.7. Trust had close to as many definitions as did the very vague terms 'love' and 'like.' Hence, trust is by nature hard to narrow down to one specific definition because of the richness of meanings the term conveys in everyday usage.

Bigley and Pierce [6] chronicled the different uses of the words trust and distrust, showing both how various definitions are similar and how they diverge. Trust conceptualizations have ranged from a personality construct [79] to a rational choice [11] to an interpersonal relationship [72] to a social structure construct [51]. Bigley and Pierce argued that because these "are not trivial differences ... efforts to incorporate existing trust perspectives under **one** conceptualization are likely to result in concepts that are either unreasonably complex or inordinately abstract for ... research purposes" ([6: 415]; emphasis added). Indeed, the differences among these trust conceptualizations appear incommensurable, indicating that researchers need a new paradigm of the meanings of trust [47].

But this is challenging, because, in the search for a new definition-of-trust paradigm, researchers must find a delicate balance. The resulting trust constructs must be:

1. Comprehensive enough to cover most of the conceptual meaning the word trust conveys in ordinary use, so that scientific work on trust will be grounded in practice [3, 43, 54] – otherwise, research results will not be useful to practice [82].
2. Not so large and complex individually that they stretch trust's conceptual meaning into vagueness [70].
3. Able to convey the original meaning from prior researchers' models across disciplines in order, where possible, to build on prior research.

If one considers trust a unitary concept or stays within strict singular disciplinary bounds, this three-fold challenge is impossible.

However, if one treats trust as a set of interdisciplinary concepts, then perhaps the challenge can be met. Using an interdisciplinary approach accords with the growing consensus that trust is not unitary, but is a multiplex of concepts [41, 57, 80]. Using a set of concepts permits broader coverage, satisfying the first requirement. Forming specific constructs within the set allows each to be conceptually focused so that the individual construct does not stretch, meeting the second requirement. Making each individual construct specifically tied to one disciplinary frame maintains the original meaning of prior researchers, meeting the third requirement. In essence, we propose researchers use the divide and conquer approach to define the mammoth we call trust: *"How do you eat an elephant?" "One bite at a time"*.

At least three other challenges exist for building a good set of trust concepts:

1. To produce trust constructs that can be measured. Defining variable-level constructs that are well defined and specific meets this challenge.
2. To connect the constructs in meaningful ways. Schwab said, "constructs are of interest only if they are connected to other constructs" [85: 6]. We connect trust with power and control in this article. Also, each construct in a typology of distinct trust constructs would differ from another, offering a chance to

connect them. Thus, a typology of trust concepts that relate to each other would be helpful. Tiryakian said, "a good typology is not a collection of undifferentiated entities but is composed of a cluster of traits which do in reality 'hang together' " [94: 178].

3. The constructs should be parsimonious enough to be easily understood and clearly distinguishable from each other conceptually.

3   Creating a Typology for Trust

We initially explored the possibilities for a trust typology by analyzing definitions in the trust literature. We found sixty-five cited articles and monographs that contained definitions of trust – twenty-three from the psychology domain, twenty-three from management or communications, and nineteen spread across sociology, economics or political science. Our analyses were relatively free from bias because we do not belong to any of these disciplines. We noticed that the trust definitions could be categorized by trust <u>referent</u>, which is typically the characteristics of the trustee (e.g., goodwill, honesty, morality, expertness, caring; cf. [57]). We categorized the characteristics in each article or monograph into sixteen logical groupings [60]. By comparing these categories with each other, four high level categories resulted: *benevolence, integrity, competence, and predictability. Benevolence* means caring and being motivated to act in one's interest rather than acting opportunistically [34]. *Integrity* means making good faith agreements, telling the truth, and fulfilling promises [9]. *Competence* means having the ability or power to do for one what one needs done [3]. *Predictability* means trustee actions (good or bad) that are consistent enough to be forecasted in a given situation [24]. Predictability is a characteristic of the trustee that may positively affect willingness to depend on the trustee regardless of other trustee attributes. In our categorizations of definitions, goodwill, responsiveness, and caring fell into the benevolence category, while honesty and morality were categorized as integrity, and expertness was classified as competence. Ninety-two percent of the definitions that involved trustee characteristics fell within these four categories.

We also noticed that the definitions could be categorized by conceptual type, such as attitude [42], intention [12], belief [36], expectancy [10], behavior [29], disposition [78], and institutional/structural [27, 86]. We combined the belief and expectancy categories, since these terms differed primarily in terms of present versus future orientation. Since categorizing by conceptual type did not overlap with categorizing by referent or characteristic, we placed referents and conceptual types on separate axes of a table. Then we mapped the definitions from the literature onto the matrix (Table 1). The result was the expected finding that literature trust definitions were almost all over the map.

This matrix provides a way to conceptually compare various trust definitions. Each "x" in Table 1 represents one trust definition. The 'Other characteristics' row includes attributes like openness and carefulness. 'Other referent' refers to either people or institutions. Beliefs and attitudes that refer to a particular characteristic are generally phrased something like "One believes (or, if an attitude,

Table 1. Mapping of Literature Trust Definitions

Trustee Charac- teristic / Referent	Conceptual Types					
	Dispo- sition	Struc- tural	Affect / Attitude	Belief / Expectancy	Intention	Behavior
Competence			x	xxxxxxxxxxxx xxxxxxxx		xxxxx
Benevolence			xxxxxxxxx xxx	Xxxxxxxxxxx xxxxxxxxxxxxx xx	Xxx	xxxxxx
Integrity			xxxxxxxx x	xxxxxxxxxxxx xxxxxx	X	xxxxxxx
Predictabil- ity			x	xxxxxxxxxxx		x
Other characteristic				xxxxxxx		x
Other Referent	xxxxx	xxxxxx	xxxxxxx	xxxxxxxxxxxx xxxxxxxxx	Xxxxx	xxxxxxxxxx xxxxx

'One feels secure') that the trustee is (e.g., competent)". Articles with marks
at the intersection of Behavior and Benevolence have definitions reflecting that
one would behaviorally depend on the other's benevolence. For example, Baier
[2: 235] said, "When I trust another, I depend on her good will toward me".
Similarly, a definition at the intersection of Intention and Competence would
say that one intends to depend on the other's benevolence (e.g., [9]). Notice that
a plurality of intention and behavior definitions referred only to the person ('I
depend on O' or 'I am willing to be vulnerable to O'), which seems more natural. We can believe that another person has benevolence. But to say we intend
to depend on that benevolence is not as precise, because it mixes a willingness
or intention with a perception about the trustee's attribute. Rather, we more
correctly say we intend to depend on the other person because (we believe) they
are benevolent, making benevolence belief more properly the antecedent of the
intention.

From this mapping, and from a conceptual analysis of how trust types relate
to each other [62], we created an interdisciplinary model of conceptual trust types
(Figure 1). The model has constructs representing five Table 1 columns. Disposition to trust represents the dispositional and institution-based trust represents
the structural. Trusting intentions and trust-related behavior represent the intention and behavior columns in Table 1. The affect/attitude and belief/expectancy
columns were combined into the construct 'Trusting Beliefs' because it is so difficult to distinguish affect/attitude and belief. This is a calculated departure
from McAllister [58], who proposed affective and cognitive trust types, based on
our empirical work. An initial analysis of the wording of McAllister's items indicated that most of the cognitive items reflected job competence, while most of
the affective items fell into the benevolence category. We selected a typical item
from each scale and had trained student raters categorize it, along with items
from other scales, into the four trusting belief categories and pure affect/liking.

Between sixty and sixty-five percent of the raters agreed that these items belonged in the competence and benevolence categories. The next highest category (affect) only received between thirteen and twenty percent of the responses. We found in an industry study (n=101) that trusting belief-competence and benevolence in one's boss were both highly correlated (above 0.70) with affect/liking items, indicating that both competence and benevolence are strongly affective in nature. Benevolence and competence beliefs also correlated highly in an Internet study. Therefore, it seemed better to use these constructs as belief constructs, incorporating affect into the definitions where possible, as did Rempel, Holmes and Zanna [72]. Dispositional and institutional trust are also shown in Figure 1, representing the disposition and structural columns of Table 1.

4   Trust Typology Conceptual Definitions

We define each trust type depicted in Figure 1 at the conceptual level. In conjunction with each high level trust type, we define measurable constructs that are subsets of each of the five trust types. As we provide definitions of the constructs in Figure 1, we discuss aspects that tend to be included in trust definitions, such as situational specificity, risky or uncertain conditions, feelings of security or confidence, and absence of a control basis for trust. Trusting intentions will be defined and described first. Links among constructs are rather intuitive. The links among trusting beliefs, trusting intentions, and trust-related behavior follow the general pattern of the theory of reasoned action [21], except that attitude/affect is included in the construct definitions rather than as a separate construct. More rationale on the links among these constructs may be found in [57] and [62] and is summarized briefly at the end of this section. To stay internally consistent, all trust constructs are defined at the individual level of analysis.

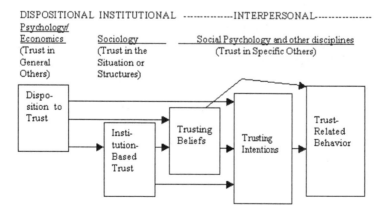

Fig. 1. Interdisciplinary model of trust constructs.

Several researchers have defined trust as an intentional construct (e.g., [12, 84]). **Trusting Intentions** means one is willing to depend, or intends to depend, on the other party with a feeling of relative security, in spite of lack of control over that party, and even though negative consequences are possible. This Trusting Intentions definition embodies four elements synthesized from the trust literature. 1. The possibility of negative consequences [26, 104] or risk [29, 73, 86] or uncertainty [66, 72] makes trust important but problematic. One who trusts is exposed to greater potential harm from a breach of trust than the expected benefit if the trustee comes through [15, 54]. 2. A readiness to depend or rely on another is central to trusting intentions [29, 51, 74]. By depending on another, one makes oneself vulnerable to the results of trustee freedom to act [57]. Freedom to act is assumed in trust relations [54]. 3. A feeling of security means one feels safe, assured, and comfortable (not anxious or fearful) about the prospect of depending on another [51, 72]. The term "relative security" means one has different degrees of felt security or confidence about being willing to depend. Feelings of security reflect the affective side of trusting intentions. To feel secure, per Webster's Ninth Collegiate Dictionary, is to feel easy in mind, confident, or assured in opinion or expectation. One Webster's definition of confidence involves being certain. Therefore, feelings of relative security involve degrees of confidence or certainty. 4. Trusting intentions involves willingness that is not based on having control or power over the other party [26, 57, 73, 74]. This part of the definition implies that trust is not based on deterrence [80]. Rather than trusting in controls, the trustor trusts in trust [26, 74]. Using control in the definition of trust helps link trust to the control literature, and provides a better conceptualization of trust, since trust and control, though separate, are integrally linked (e.g., [12]).

Our definition refers to willingness or intention to depend on the other person generally–not in a specific situation, as some have proposed [24, 90]. This makes the other *person* the object of trust, rather than the person in one situation. Although it is perfectly valid to think of trust as situation-specific (and it is often measured that way), we chose to define our interpersonal trust constructs to refer to the person her/himself in order to provide an overall picture of the relationship between the trustor and the trustee. Hence, researchers who further decompose trust constructs into particular trust-related situation segments would obtain indicators of the overall relationship between trustor and trustee.

Trusting intentions subconstructs include Willingness to Depend and Subjective Probability of Depending. Willingness to Depend means one is volitionally prepared to make oneself vulnerable to the other person by relying on them, with a feeling of relative security [57]. Subjective Probability of Depending means the extent to which one forecasts or predicts that one will depend on the other person, with a feeling of relative security [12]. Subjective Probability of Depending is more specific and indicates a firmer commitment to depend than does Willingness to Depend.

Trust-related Behavior means that a person voluntarily [50] depends on another person with a feeling of relative security, even though negative consequences are possible. 'Depends' is specified as a behavioral term, distinguishing trust-related behavior from trusting intentions, which inhere a willingness to depend. A number of researchers have defined trust as a behavior (e.g., [2, 7, 29, 73]). Trust-related behavior means one gives another person a fiduciary obligation [3] by acting such that the other could betray them. Thus, Trust-related Behavior implies acceptance of risk, just as Mayer et al. [57] argued.

When a trustor behaviorally depends on a trustee, it gives the trustee some measure of power over the trustor, because dependence is the obverse of power [19]. One who depends on the other person places the other in a position of dependence-based power over one. Therefore, by definition, behaviorally trusting another voluntarily gives the other dependence-based power over them. While the trust literature has discussed the interplay between trust and power (e.g., [7, 22, 24, 91]), trust and power have not previously been linked by definition. Linking these terms by definition should help integrate the trust and power literatures, which is important, per Kaplan [40: 321]. Helpful links may be made via such power definitions as Walton's [97], that dependence means one's interest (what is at stake) in satisfactions provided by the other party.

Trust-related behavior comes in a number of subconstruct forms because many actions can make one dependent on another. As exemplar constructs, we outline (rather than define) the following trust-related behaviors here: cooperation, information sharing, informal agreements, decreasing controls, accepting influence, granting autonomy, and transacting business. Each trust-related behavior makes the trustor dependent upon the trustee. Cooperation with another makes one dependent on the other not to betray, for instance, in a prisoners dilemma situation [14, 91]. Cooperating instead of competing with another (e.g., on a research paper) also opens up the possibility of having to share rewards and makes one dependent for success on the actions of another agent. Information sharing (e.g., [66]) represents trust-related behavior because it makes one vulnerable to the actions of the trustee with respect to the information. For example, giving your social security or credit card number to an online vendor makes one vulnerable to the sale or illegal use of that information. One has to depend on them to keep your numbers secure and private. When one enters an informal agreement with another [12, 86], one depends on the other party to keep its part of the agreement without the benefit of legal contract enforcement. By reducing the controls or rules we place on another person [2, 22], we are exposing ourselves to more risk because we have to depend on that other person without being able to direct them or to detect breaches of trust. When we allow someone to influence us [7], we depend on their opinion being correct, because if it is incorrect, negative consequences may follow. Similarly, when a boss grants an employee more decision-making power [10], the boss must depend on the employee to make the right decisions. When one transacts business with an Internet vendor, one depends on the vendor to deliver the goods/services and to keep one's personal information confidential. Every trust-related behavior occurs under risk and either the inability, or lack of desire, to control the trustee.

Trusting beliefs are cognitive perceptions about the attributes or characteristics of the trustee. Often, people trust behaviorally because of inferences about the trustee's traits [103]. A number of researchers have defined trust as a cognitive belief or expectancy [9, 24, 31]. **Trusting Beliefs** means the extent to which one believes, with feelings of relative security, that the other person has characteristics beneficial to one. One judges the trustee to be trustworthy, meaning that they are *willing and able* to act in the trustor's interest [17, 64]. Like trusting intentions and trust-related behavior, trusting beliefs is defined to be person-specific but not situation-specific.

Based on the above-mentioned analysis of the types of trusting beliefs, we define four subconstructs, which we illustrate with the relationship between a consumer and an Internet vendor. Trusting Belief-Competence means one securely believes the other person has the ability or power to do for one what one needs done. In the case of the Internet relationship, the consumer would believe that the vendor can provide the goods and services in a proper and convenient way. Trusting Belief-Benevolence means one securely believes the other person cares about one and is motivated to act in one's interest. A benevolent Internet vendor would not be perceived to be apt to act opportunistically. Trusting Belief-Integrity means one securely believes the other person makes good faith agreements, tells the truth, and fulfills promises [9]. This would reflect the belief that the Internet vendor will come through on his/her promises, such as to deliver goods or services or to keep private information secure. Trusting Belief-Predictability means one securely believes the other person's actions (good or bad) are consistent enough that one can forecast them. People with high Trusting Belief-Predictability would believe that they can predict the Internet vendor's future behavior. This construct, as opposed to Trusting Belief-Integrity, is value-neutral, such that the vendor is believed predictably to do either good or bad things in the future.

These four trusting beliefs together provide a strong foundation for trusting intentions in the other party and at the same time fulfill our "willing and able" definition of a trustworthy trustee. If the trustee is benevolent, he/she is *willing* to help us. However, just being willing is not sufficient because they may not be *able* to help. The competent trustee is able to help us, so these two together are much more powerful than either is alone. Even so, the trustee may not come through on an agreed-upon action, as we would like. However, one who has integrity will prove a willingness to help by making and fulfilling good faith agreements with us. But is the trustee going to come through every time, or do we still have to worry? A trustee who is predictable will not vary or change from time to time. Therefore, the combination of the four trusting beliefs provides a firm foundation for trusting intentions and trust-related behavior [57, 61]. A trustee who is consistently (predictable) shown to be willing (benevolent) and able (competent) to serve the trustor's interest in an honest, ethical manner (integrity) is indeed worthy of trust. Of some people, we perceive that they have all four qualities. Of others, we perceive that they are strong in one characteristic, but weak in another. Which characteristic is most important depends on the context [63]. Perhaps the diversity of trustees and trustee attributes is what makes trust an interesting proposition!

So far we have defined terms related to interpersonal trust (Figure 1). That is, we trust other people, either personally, as in trusting behavior and trusting intentions, or their attributes, as in trusting beliefs. However, the object of trust may involve situations and structures instead of people [22, 27]. **Institution-based Trust** means one believes, with feelings of relative security, that favorable conditions are in place that are conducive to situational success in a risky endeavor or aspect of one's life [50, 54, 86, 105]. This construct comes from the sociology tradition positing that people can rely on others because of structures, situations, or roles [2] that provide assurances that things will go well. Zucker [105] traced the history of regulations and institutions in America that enabled people to trust each other-not because they knew each other personally, but because licensing or auditing or laws or governmental enforcement bodies were in place to make sure the other person was either afraid to harm them or punished if they did harm them. Institution-based trust refers to beliefs about those protective structures, not about the people involved. Therefore, it focuses on an impersonal object. However, institution-based trust affects interpersonal trust (Figure 1) by making the trustor feel more comfortable about trusting others in the situation.

Institution-based Trust has two subconstructs, Structural Assurance and Situational Normality, which come from two separate sociological traditions. Structural Assurance means one securely believes that protective structures– guarantees, contracts, regulations, promises, legal recourse, processes, or procedures – are in place that are conducive to situational success [86, 100, 105]. Structural assurance reflects the idea that trusting intentions are set up or structured environmentally. That is, structural conditions amenable to trusting intentions build trusting intentions. For example, one using the Internet would have structural assurance to the extent that one believed legal and technological safeguards (e.g., encryption) protect one from privacy loss or credit card fraud. People believe in the efficacy of a bank to take care of their money because of laws and institutions like the Federal Deposit Insurance Corporation (FDIC) that assure against loss. With a high structural assurance level, one would be more likely to be willing to rely on a specific bank because of the secure feeling structural assurance engenders. In organizations, Structural assurance might refer to the processes and procedures that make things safe or fair in that specific organizational setting. An example of a structural assurance is seen in the 1986 Challenger space shuttle disaster. Starbuck and Milliken [92] said that successful organizations come to be confident in their assurance procedures. "They trust the procedures to keep them appraised of developing problems . . . " [92: 329-330]. On the day before the disastrous shuttle launch, after twenty-four successful shuttle launches, Mulloy, one of NASA's managers, objected to using cold weather as a Launch Commit Criteria. Mulloy pointed out that the existing Launch Commit Criteria had always worked in the past. "Mulloy spoke as if he had come to trust the Launch Commit Criteria that had always produced successes" [92:

330]. These criteria had come to be perceived as a structural assurance that a launch would succeed.

Situational Normality means one securely believes that the situation in a risky venture is normal or favorable or conducive to situational success. Situational normality reflects Garfinkel's [27] idea that trust is the perception that things in the situation are normal, proper, customary [3], fitting, or in proper order [50]. Garfinkel found in natural experiments that people don't trust others when things "go weird", that is, when they face inexplicable, abnormal situations – because the situation itself is untrustworthy. For example, one subject told Garfinkel's experimenter that he had a flat tire on the way to work. The experimenter responded, "What do you mean, you had a flat tire?" The subject replied, in a hostile way, "What do you mean? What do you mean? A flat tire is a flat tire. That is what I meant. Nothing special. What a crazy question!" [27: 221]. At this point, interpersonal trust between them broke down because the illogical question produced an abnormal situation (infecting the subject with low situational normality). High situational normality means one perceives that a properly ordered setting exists that is likely to facilitate a successful venture. When one believes one's role and others' roles in the situation are appropriate and conducive to success, then one has a basis for trusting the people in the situation. Hence, situational normality is likely related to Trusting Beliefs and Trusting Intentions. An employee who feels good about the roles and setting in which he or she works is likely to have Trusting Beliefs about the people in that setting.

Our definitions represent the impersonal focus of institution-based trust as a belief held by an individual about impersonal things (the underlying structures and situations). While some sociologists cringe at the use of an individual cognitive focus [51], other sociologists (e.g., [3]) have used cognitive definitions in order to clarify the conceptual meaning of a construct for use in explaining a social phenomenon. Situating institution-based trust as a mental concept also makes it consistent with the mental constructs Trusting Beliefs and Trusting Intentions. In this way, the typology stays internally consistent.

A number of researchers have studied trust as a dispositional variable (e.g., [20, 76, 78, 83, 96]). **Disposition to Trust** means the extent to which one displays a consistent tendency to be willing to depend on general others across a broad spectrum of situations and persons. Disposition to trust differs from trusting intentions in that it refers to general other people rather than to specific other people. This construct hails primarily from dispositional psychology. Our definition does not literally refer to a person's trait. Rather, it means that one has a general propensity to be willing to depend on others [57]. As an example, one employee we interviewed, when asked whether he trusted his new boss, replied that he generally trusts new people, both at work and elsewhere. Disposition to trust does not necessarily imply that one believes specific others to be trustworthy. Whatever the reason, one tends to be willing to depend on others generally. People develop Disposition to Trust as they grow up [20], though it is altered by experiences later in life. It is a generalized reaction to life's experiences with other people [78]. Because Disposition to Trust is a generalized tendency across situations and persons, it probably colors our interpretation of situations and actors in situations, but only has a major effect on one's trust-related behavior when novel situations arise, in which the person and situation are unfamiliar [39].

Disposition to Trust has two subconstructs, Faith in Humanity and Trusting Stance. Faith in Humanity refers to underlying assumptions about people, while Trusting Stance is like a personal strategy. Faith in Humanity, from psychology, means one assumes general others are usually honest, benevolent, competent, and predictable (e.g., [76, 102]). Faith in humanity differs from trusting beliefs in that it refers to general others, while trusting beliefs refers to specific other people. Mayer et al. [57] gave the example that if you were going to drown, could you trust that strangers in the area would come to your aid? You would if, having high Faith in Humanity, you assumed others generally care enough to help.

Trusting Stance means that, regardless of what one assumes about other people generally, one assumes that one will achieve better outcomes by dealing with people as though they are well-meaning and reliable. Therefore, it is like a personal choice or strategy to trust others. Luhmann [54] said this might occur as one considers how essential trust is to one's ability to function in the social world. Because it involves choice that is presumably based on subjective calculation of the odds of success in a venture, Trusting Stance derives from the calculative, economics-based trust research stream (e.g., [73]). Here's an example. We once asked an IS employee if she trusted her newly hired manager, whom she had never met before. She said that she did trust her, because she always trusted new people until they gave her some reason not to trust them. Thus, she had a high level of Trusting Stance, which encouraged her to be willing to depend on her new boss.

Trusting Stance and Faith in Humanity are alike in that they each constitute a tendency or propensity [57] to trust other people. They differ in terms of the assumptions on which they are built. Because Faith in Humanity relates to assumptions about peoples' attributes, it is more likely to be an antecedent of Trusting Beliefs (in people) than is Trusting Stance. Trusting Stance may relate more to Trusting intentions, which may not be based wholly on beliefs about the other person [62].

We now further justify the links in Figure 1. Trust-related behavior is directly caused by trusting intentions and trusting beliefs because people tend to translate their beliefs and intentions into actions [21]. Significant support exists for the effects of trusting intentions or trusting beliefs on trust-related behaviors like information sharing (e.g., [33, 68]) and cooperation (e.g., [8, 53]). While fewer studies support links from intentions and beliefs to other trust-related behaviors (e.g., [12]), it makes sense that any behavior that increases one's vulnerability to another would be encouraged by intentional or cognitive trust in that person. Per the theory of reasoned action [21], trusting intentions will partially mediate the effects of trusting beliefs on trust-related behavior. Institution-based trust is a condition for interpersonal trust because trust in the situation typically leads one to trust in the people within the context [62]. Hence, Figure 1 depicts institution-based trust as a factor of both trusting beliefs and trusting intentions. Disposition to trust will relate positively to institution-based trust because what one believes about others generally should rub off on what one believes about the institutions or structures in which other people are involved. Disposition to trust will also influence trusting beliefs and trusting intentions, but that influence will be almost fully mediated by institution-based trust except in novel situations [39, 78]. We have gathered empirical data consistently supporting the links among the trust typology constructs.

5   Potential Trust Typology Model Extensions

In Figure 1, we can see three very different types of trust–dispositional, institutional, and interpersonal. One way to distinguish among these three types is by their disciplinary sources, as Figure 1 indicates. Another way is to consider them within a 'grammar of trust'. In all our definitions, trust has been treated as an action verb. One trusts the trustee. But action verbs like trust have both subjects and direct objects. Without making it obvious as a part of speech, we have actually discussed the subjects and direct objects of trust. The term "trustor" is the subject of the verb trust and the term "trustee" is the direct object. With all three types of trust, the trustor is the same – an individual. That is, all our definitions have used the individual level of analysis, making trust something seen through one person's eyes. However, the trustee differs by trust type. With interpersonal trust types, the direct object is the other specific individual. The trustor trusts the specific trustee. With disposition to trust, the direct object is people in general. The trustor trusts others generally. With institution-based trust, the direct object is the environmental structures or situation: the trustor trusts the structure or situation.

With this distinction among types of trust, we can extend the model another step. Within the interpersonal trust category, we had trusting beliefs that referred to the attributes of a specific person. Since the disposition to trust construct we called 'faith in humanity' is also about persons, we can specify characteristics for them as well. Just as a trustor can believe in the benevolence of a specific person (trusting belief-benevolence), so the trustor can believe in the benevolence of people generally. We would call this subconstruct 'faith in humanity-benevolence', meaning that one assumes that general others are usually benevolent. Applying the same principle, we obtain the constructs 'faith in humanity-competence,' '-integrity,' and '-predictability'.

Can this same principle be applied to institution-based trust? It may, to the extent that structures or situations may be thought to have attributes. In some cases, this is inappropriate because it reifies inanimate objects; but in other cases, it makes sense to apply attributes to institutions. For instance, an environment that has safeguards like credit cards with their limits of $50 fraudulent use loss maximums might be considered benevolent-not in terms of moral intentions but in terms of safe conditions. Or, if law enforcement were extremely thorough and effective in an environment like banking, then we might say the banking environment demonstrates integrity. If a norm or capability develops among Internet booksellers to always provide fast shipment, this would constitute an area of competence in the Internet book-purchasing situation.

Further, the full extent of the possibilities for the model's trust constructs can be conceived as a $3 \times 3 \times 4$ matrix that defines the combinations of the object of trust (represented by dispositional, institutional, interpersonal), the construct type (belief, intention, and behavior), and the attribute of the trustee (benevolence, integrity, competence, predictability). The constructs defined in section 4 cover a portion of these. Some additional ones make sense. Intentions could appropriately be applied to faith in humanity, for example, because one can be willing to depend, or intend to depend, on other people generally. Further, one can behaviorally depend on others generally, making behaviors useful in the dispositional domain. Likewise, one can be willing to depend and behaviorally depend on the situations and structures of the institution. On the other hand, assigning trustee characteristics to the intentional and behavioral constructs does not make sense. Characteristics only apply to the belief-like constructs. This is because intentions and behaviors involve doing (or intending to do) something, whereas beliefs involve perceptions of states of being.

6   The Nature of Distrust

A number of researchers have discussed or defined distrust. Deutsch [15] said that a distrusting choice is avoiding an ambiguous path that has greater possible negative consequences than positive consequences, a definition that is the opposite of his definition of trust. Early on, Deutsch [14] used the term "suspicion" for distrust. Using either term, it is clear that Deutsch felt distrust and trust were opposites. After examining Webster's and Random House dictionary definitions of distrust, suspicion, suspect, and doubt, we concluded that these terms differ only in degree, not in kind. Webster's defines distrust as the absence of trust (synonyms – suspicion, wariness), whereas it defines suspicion as to suspect something is wrong, to mistrust, or to doubt. The only difference between suspicion and distrust seems to be that suspicion may be based on slight evidence, while evidence is not mentioned in dictionary definitions of distrust. Fox also refers to distrust (either personal or institutionalized) as the opposite of trust: "I trust my friends; distrust my enemies" [22: 67].

Distrust and trust can apparently reside in the same person at the same time. For example, Gellner [28] said that anarchy, a state that reflects distrust among many people) is what engenders trust among a few people trying to band together to counteract anarchy. Sitkin and Roth [90] differentiated trust and distrust as two very distinct constructs. Dunn [18: 74] quoted Hobbes as saying that while trust is a passion proceeding from the belief of one from whom we hope something good, distrust is "diffidence or doubt that makes one try to find other means". Trust and distrust could co-exist because conceivably, one could hope for something good at the same time look for backup means to get that same good thing. On the other hand, Worchel [101] said that trust and mistrust are two extremes of the same dimension, as did Rotter [79]. "Mistrust", per Worchel, is "a sense of readiness for danger and an anticipation of discomfort" [101: 176].

Distrust is not only the opposite of trust, but "also a functional equivalent for trust" (Luhmann [54: 71]). One chooses between the two. While trust reduces (and thus solves the problem of) the complexity of the social system, distrust by itself does not. Hence, the untrusting must use other strategies to reduce complexity. Luhmann [54] identifies these as including the definition of one's partner as the enemy, the building up of huge emergency reserves, and the renunciation of one's needs. "These negative strategies give distrust that emotionally tense and often frantic character which distinguishes it from trust ... Strategies of trust become correspondingly more difficult and more burdensome". [54: 71-72]. In terms of emotion, then, one might picture trust as the satisfied zoo elephant, calmly eating hay, while distrust is more like the raging wild bull elephant charging the tusk hunter who threatens the herd. Other complexity-reducing distrust strategies include placing controls over the trustee.

Although some have said that trust and distrust are two ends of the same conceptual spectrum (e.g., [79]) most trust theorists now agree that trust and distrust are separate constructs that are the opposites of each other [49, 90]. If distrust were the same as low trust, this would imply that trust and distrust are opposite levels of the same construct. But distrust is apparently more like the opposite of trust; hence, both can have high or low levels. Barber [3: 166] defined distrust as the opposite of trust: "rationally based expectations that technically competent performance and/or fiduciary obligation and responsibility will *not* be forthcoming". Only the word "not" distinguishes Barber's definition of distrust from his definition of trust.

Similarly, dictionary definitions of trust and distrust indicate that these are opposites. Webster's Ninth Collegiate Dictionary says that distrust is "the lack or absence of trust: suspicion, wariness". As a verb, it defines distrust as "to have no trust or confidence in". Webster's also uses the term 'confidence' in its verb and noun definitions of trust. The unabridged Random House dictionary says distrust, as a verb, is "To regard with doubt or suspicion; have no trust in". As a noun, distrust is "lack of trust; doubt; suspicion". These definitions all imply that distrust is the opposite of trust.

Lewicki and associates [49], who have probably done the most thorough contrast of trust and distrust to date, argued that trust and distrust are separate for three reasons: a) they separate empirically, b) they coexist, and c) they have different antecedents and consequents. As empirical evidence that the two concepts are separate, Lewicki et al. [49] cited Wrightsman [102], whose philosophies of human nature scale separated into two factors, one with positively worded items (disposition to trust) and the other with negatively worded items (disposition to distrust). To show that trust and distrust coexist, Lewicki et al. used Mancini's [56] fieldwork to show that politicians and journalists both trust and distrust each other. To show that the antecedents and consequents of trust and distrust probably differ, Lewicki and associates reviewed the analogous literature on positive/negative affectivity, which has evidence that antecedents/consequents differ by positive/negative constructs. Our own empirical work has shown that dispositional trust and distrust constructs act differently as antecedents of other variables.

While explaining that trust and distrust are separate, Lewicki and associates also argue that trust and distrust are conceptual opposites. They provided definitions of trust and distrust that used basically the same terms. Trust is "confident *positive* expectations regarding another's conduct", while distrust is "confident *negative* expectations regarding another's conduct". They position their definitions of both trust and distrust as reflecting "movements toward certainty" [49: 439].

7 Distrust Typology Conceptual Definitions

Because so many researchers have proposed distrust as a construct defined as the opposite of trust, we felt secure in forming distrust definitions that are the mirror image of our trust definitions. Although far fewer distrust studies have been conducted than trust studies, some of the constructs defined below build on prior literature (e.g., [90, 101]). In one set of instances, based on how the literature describes and defines distrust and related words (suspicion, doubt), we use words in the definitions that are similar, but differ slightly from those used for trust definitions. Whereas we used the term 'feelings of security' in our trust definitions, we use the terms 'certainty' or 'confidence' in describing distrust constructs, since these terms are used much more frequently in definitions of distrust than is the term 'insecurity.'

What follows are the distrust typology definitions. We include distrust definitions corresponding to all the constructs defined in the Section 4 trust typology. **Distrusting Intentions** means one is not willing to depend, or intends not to depend, on the other party, with a feeling of relative certainty or confidence, even though negative consequences are possible. The feeling of relative certainty or confidence refers to the intention not to depend, not to the other party. That is, one feels relatively certain or confident in one's intention not to depend. Two distrusting intentions subconstructs are now defined. No Willingness to Depend means one is not volitionally prepared to make oneself vulnerable to the other person by relying on them, with a feeling of relative certainty or confidence. Subjective Probability of Not Depending means the extent to which one forecasts or predicts that one will not depend on the other person, with a feeling of relative certainty or confidence.

Distrust-related Behavior means that a person does not voluntarily depend on another person, with a feeling of relative certainty or confidence, when negative consequences are possible. The following distrust-related behaviors are outlined: lack of cooperation, information distortion, formal agreements, increasing controls, not accept influence, not grant autonomy, and no business transacting.

Each of these has in common a need to reduce dependence on the other person or to "minimize any potential damages that may result from having to trust others" [44: 42]. Lack of cooperation is a typical outcome of distrusting intentions in an organization [9] because distrust paralyses capacity for cooperative agency, per Dunn [18] and may even lead to sabotage, based on Zand's [104] findings. Distrusting intentions lead to information distortion, per Bromiley and Cummings [9]. Similarly, McGregor [59] said that without trusting intention, the openness of communication is limited. Others have also found empirically that distrusting intentions lead to withheld or distorted information [25, 65, 69, 75, 104] or even to deception [14, 45]. Instead of allowing agreements to become less formal, those with low trusting intention tend to want to formalize their agreements, in order to be able to apply legal processes to the arrangement in case of a breakdown [89]. Likewise, the distrustor desires more control over the trustee, since s/he cannot trust trust. Zand [104] describes how one with distrusting intention requires increasing control to assure that things will go right. Increasing controls would include behavior monitoring (e.g., [12]). Those with distrusting intention do not accept influence from the other [7, 68] because they are suspicious of the other's motives. We do not grant them as much autonomy [22, 44] because we want to limit the wrong they may do on our behalf. The ultimate weapon of one with distrusting intention in a market relation is to do no business transacting with the trustee [87]. This eliminates dependence on them completely. In addition to the above distrusting behaviors, the dark side of trust includes negative behaviors like whistle-blowing, feuding, revenge, and even violence [5] that we only mention in passing.

Distrusting Beliefs means the extent to which one believes, with feelings of relative certainty or confidence, that the other person does not have characteristics beneficial to one. Four specific distrusting beliefs are now defined. Distrusting Belief-Competence means one, with some degree of confidence, believes the other person does not have the ability or power to do for one what one needs done. Distrusting Belief-Benevolence means one, with some degree of confidence, believes the other person does not care about one and is not motivated to act in one's interest. Distrusting Belief-Integrity means that, with some degree of confidence, one believes the other person does not make good faith agreements, does not tell the truth, and does not fulfill promises. Distrusting Belief-Predictability means that, with some degree of confidence, one believes the other person's actions (good or bad) are not consistent enough that one can forecast them in a given context.

Lewicki et al. [49] indicate that one may have both high interpersonal trust and high interpersonal distrust because people trust each other in one situation but not in another. As one gets to know the other person, the relationship becomes multi-faceted because of both negative and positive experiences with the other person. We believe this is true because the strength of confidence one has in the other person will vary from situation to situation. For example, a trustor who would normally trust a trustee may distrust the trustee when the trustee has strong incentives to defect.

High trust and distrust would not simultaneously exist if interpersonal trust constructs were defined as situation-specific, because it is difficult to imagine both highly trusting and highly distrusting a person regarding the same situation. For example, if one highly trusts a marriage partner to be faithful given a particular temptation, one cannot highly distrust them to be faithful in the same condition. Therefore, situation-specific definitions do not facilitate the simultaneous existence of high trust and distrust. The same would be true of simultaneously low trust and distrust.

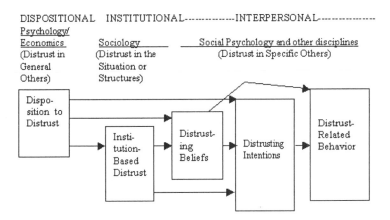

Fig. 2. Interdisciplinary model of distrust constructs.

Institution-based Distrust means one believes, with feelings of relative certainty or confidence, that favorable conditions that are conducive to situational success in a risky endeavor or aspect of one's life are not in place. No Structural Assurance means one confidently believes that protective structures that are conducive to situational success–guarantees, contracts, regulations, promises, legal recourse, processes, or procedures–are not in place. No Situational Normality means one confidently believes that the situation in a risky venture is not normal or favorable or conducive to situational success.

Disposition to Distrust means the extent to which one displays a consistent tendency to not be willing to depend on general others across a broad spectrum of situations and persons. Suspicion of Humanity means one assumes general others are not usually honest, benevolent, competent, and predictable. Suspicion of Humanity subconstructs are assumed to involve the four attributes described with our earlier definitions for Faith in Humanity. Distrusting Stance means that, regardless of what one assumes about other people generally, one assumes that one will achieve better outcomes by dealing with people as though they are not well-meaning and reliable.

The distrust constructs should interrelate in the same manner in which the trust constructs relate (Figure 2). Disposition to distrust, for example, should be positively related to institution-based distrust because suspicions and doubt about other people generally should be closely aligned with suspicions and doubts about a specific environment in which people operate. In turn, institution-based distrust should be positively related to interpersonal distrust in the specific people in the situation.

On the other hand, we are less sure that distrust constructs will relate strongly to trust constructs. Although distrust, like trust, involves strong affect, it may or not be related to the affective side of trust. Since distrust constructs often reflect such emotions as wariness, caution, cynicism, defensiveness, anger, fear, hate, and a feeling of betrayal [49, 98], as well as uncertainty and lack of confidence, distrust tends to differ in its makeup from trust, which reflects such emotions as hope, safety, assurance, and confidence. These sets of emotions may be more orthogonal than merely at different ends of the same scale from each other.

8   Importance of the Trust versus Distrust Distinction

One with high interpersonal trust and low interpersonal distrust would tend to ignore or rationalize away evidence that the other party is not trustworthy [34]. For example, an accounting firm partner who has developed high trust in the client firm may have trouble seeing evidence of malfeasance even when it is presented. Lewicki et al. [49] cited evidence for this and explained that an unhealthy low level of distrust may accompany high trust because little behavior monitoring takes place, resulting in greater chance of undetected fraud. This occurs because, just as love is blind, one who strongly trusts may become blind to negative aspects of the other party. But when a healthy dose of distrust enters, then one is more watchful or attentive to problems. Attention makes one more likely to find valid problems, such that they can be solved. Therefore, a balance of trust and distrust is important.

Similarly, one with low interpersonal trust and high interpersonal distrust does not have a good balance, because this combination can cause a too-jaded view of information about the other party. "Paranoid cognitions can emerge", [49: 451], such that no matter what the other party does or says, their actions and words are interpreted negatively. Many examples of this phenomenon could be cited, but it often occurs when a power differential exists between parties [46]. Hence, the parties cannot reconcile with each other even to the point of starting to cooperate.

Trust and distrust come from different sides of the personality and each finds its basis in a different concept of human nature. Trust, from the positive side, assumes the best of other people and of human institutions, a "theory Y" view of people [59]. Distrust assumes that people are opportunistic and dishonest and must be controlled. Many economic models, such as agency theory [38] and transaction cost economics [99] are based firmly on this model, the "theory X" view. We believe that trustees manifest both sides, and that each of us is capable of viewing from either side. It is often helpful to do so. Whereas looking at only the positive (trust) side of things can result in such detrimental thought patterns as "group-think" [37, 49], looking at only the negative (distrust) side can paralyze action [9, 18]. Perhaps this is because reality consists of some positives and some negatives in almost every case. Analysis of both advantages and disadvantages of an alternative is therefore important to balanced decision-making. Perhaps the 1986 Challenger shuttle mission would not have been sent to its doom if decision-makers had included more distrust in their thinking [92].

In the arena of global politics, the balanced approach is important as well. To gain cooperation among nations requires that trust be extended. Only through trusting each other can cooperative progress be made. However, to assure protection requires that nations "trust, but verify". Or, altering slightly an old New England proverb [98], "Trust [the other nation], but keep your powder dry". Trust is important for international cooperative endeavors, but neglect of distrust is not appropriate when national security or global peace are at stake.

9   Reasons the Models May Be Useful

For researchers, the need to understand both trust and distrust is patent. After years of speculation about whether these concepts are the same or different, one need only apply such definitions as these to empirical formulations to make a test. Perhaps it has been the lack of good definitions that has impeded empirical distinction between trust and distrust. In any case, most researchers have chosen to emphasize one or the other, and this may have precluded full understanding of the trust/distrust phenomenon. Most of Luhmann's [54] propositions about the interplay between trust and distrust, for example, have still not been adequately tested. If the definitions of trust and distrust constructs suggested here are used, then research results may more easily be compared, encouraging faster progress in understanding trust and distrust.

In addition, the trust/distrust models may be useful because of the following:

1. All the subconstructs are measurable with multiple items, facilitating future research (Table 2). The concepts are also amenable to interview or observation data collection methods. Although the constructs were defined at the individual level of analysis, measures can be aggregated to higher levels of analysis.
2. The constructs relate to each other in new ways that open additional research possibilities–both within the major headings of trust and distrust, and between trust and distrust. For example, the use of institution-based trust as a mediator between disposition to trust and interpersonal trust provides a new way to assess the usefulness of disposition to trust.
3. The constructs are well-defined and parsimonious enough to be easily understood and distinguished.
4. Based on an extensive literature review, these constructs not only tie to the literature, but cover the most oft-used types of trust and the key referents of trust in the literature. The definitions capture both affective and cognitive aspects of each construct.

Table 2. Trust Construct Definitions

			Interpersonal		
	Dispositional	Structural	Perceptual	Intentional	Behavioral
Trust:					
Conceptual Level	Disposition to Trust	Institution-based Trust	Trusting Beliefs	Trusting Intentions	Trust-related Behavior
Operational Level	• Faith in Humanity • Trusting Stance	• Structural Assurance • Situational Normality	Trusting Belief- • Competence • Benevolence • Integrity • Predictability	• Willingness to Depend • Subjective Probability of Depending	• Cooperation • Information Sharing • Informal Agreements • Decreasing Controls • Accepting Influence • Granting Autonomy • Transacting Business
Distrust:					
Conceptual Level	Disposition to Distrust	Institution-based Distrust	Distrusting Beliefs	Distrusting Intentions	Distrust-related Behavior
Operational Level	• Suspicion of Humanity • Distrusting Stance	• No Structural Assurance • No Situational Normality	Distrusting Belief- • Competence • Benevolence • Integrity • Predictability	• No Willingness to Depend • Subjective Probability of Not Depending	•Lack of Cooperation • Information Distortion • Formal Agreements • Increasing Controls • Not Accept Influence • Not Grant Autonomy • No Business Transacting

5. The constructs represent conceptualizations from several disciplines. Thus, they capture significant conceptual meaning from each discipline while presenting an internally consistent view of trust.

This typology compares well with most other typologies in terms of coverage. Gabarro [24], Rempel, et al. [72], and Mishra [66] only addressed trusting beliefs. Bromiley and Cummings [9] had three types each of beliefs, intentions, and behavior. Mayer, Davis and Schoorman [57] have several constructs, but their model only has three of the four trusting beliefs and no institution-based trust constructs. To our knowledge, distrust has only been delineated into a typology of constructs here.

10   Conclusion

Lewis and Weigert [50] called trust a highly complex and multi-dimensional phenomenon. Like an elephant, trust is so large that it needs to be digested a bite at a time in order to make orderly progress. Researchers should agree on what trust types exist because common definitions will enable researchers to sort out findings across studies [30]. Without agreed-upon definitions, effective meta-analyses are difficult. A recent search in ABI Inform yielded only two meta-analyses about trust, both published recently and both focused on sales relations. This meager result may be a symptom of the difficulty of comparing trust studies. Trust research needs a set of rules to interpret one result against another, as Rubin [81] recommended for the similarly vague love concept. Our delineation of the trust concept facilitates such meta-analyses by providing a way to categorize studies by trust referent and conceptual type (Table 1). When studies can be compared, consensus knowledge about trust will then progress more rapidly. This chapter's typology of trust constructs helps address conceptual confusion by representing trust as a coherent set of conceptual constructs and measurable subconstructs. One benefit of this depiction of trust is that it has heuristic value [40] by generating research possibilities. We believe the model will help researchers examine various relationships in new ways, since the model posits ways in which dispositional, institutional, and interpersonal trust and distrust concepts inter-relate. These concepts can also be modified for use with agents or between people and agents.

The trust and distrust typologies should also aid practice. Consistent definitions provide clearer means for researchers to communicate with practitioners and provide them better trust prescriptions. This dialogue would both enable trust research to be more valuable to practitioners and provide researchers the value of intuitive practitioner knowledge. Researchers like social psychologist Harold Kelley [43] have commented that the interplay between common-sense concepts and scientific concepts is useful to all. Our typologies present a vocabulary of specifically-defined trust types by which scholars and practitioners can converse on this important topic. Further, having very specific constructs should help researchers develop more specific, workable theories, and nothing is more practical than a theory that works [95]. Over time, this typology should aid development of more specific, and thus more beneficial, trust prescriptions.

References

1. Arrow, K. J.: The Limits of Organization. Norton, New York (1974).
2. Baier, A.: Trust and Antitrust. Ethics **96** (1986) 231-260.
3. Barber, B.: The Logic and Limits of Trust. New Rutgers University Press, Brunswick, NJ, (1983).
4. Barker, J. A.: Discovering the Future: The Business of Paradigms. 3rd edn. ILI Press, St. Paul, MN (1989).
5. Bies, R. J., Tripp, T. M.: Beyond Distrust: "Getting Even" and the Need for Revenge. In: Kramer, R. M., Tyler, T. R. (eds.): Trust in Organizations: Frontiers of Theory and Research, Sage, Thousand Oaks, CA (1996) 246-260.
6. Bigley, G. A., Pearce, J. L.: Straining for Shared Meaning in Organization Science: Problems of Trust and Distrust. Academy of Management Review **23** (1998) 405-421.

7. Bonoma, T. V.: Conflict, Cooperation, and Trust in Three Power Systems. Behavioral Science **21** (1976) 499-514.

8. Boyle, R., Bonacich, P.: The Development of Trust and Mistrust in Mixed-Motive Games. Sociometry **33** (1970) 123-139.

9. Bromiley, P., Cummings, L. L.: Transactions Costs in Organizations with Trust. In: Bies, R., Sheppard, B., Lewicki, R. (eds.): Research on Negotiations in Organizations, Vol. 5. JAI, Greenwich, CT (1995) 219-247.

10. Carnevale, D. G., Wechsler, B.: Trust in the Public Sector: Individual and Organizational Determinants. Administration & Society **23** (1992) 471-494.

11. Coleman, J. S.: Foundations of Social Theory. Harvard University Press, Cambridge, MA and London (1990).

12. Currall, S. C., Judge, T. A.: Measuring Trust Between Organizational Boundary Role Persons. Organizational Behavior and Human Decision Processes **64** (1995) 151-170.

13. Dasgupta, P.: Trust as a Commodity. In: Gambetta, D. (ed.): Trust: Making and Breaking Cooperative Relations Blackwell, New York (1988) 47-72.

14. Deutsch, M.: Trust and Suspicion. Journal of Conflict Resolution **2** (1958) 265-279.

15. Deutsch, M.: The Resolution of Conflict: Constructive and Destructive Processes. Yale University Press, New Haven, CN (1973).

16. Doney, P. M., Cannon, J. P., Mullen, M. R.: Understanding the Influence of National Culture on the Development of Trust. Academy of Management Review **23** (1998) 601-620 .

17. Driver, M. J., Russell, G., Cafferty, T., Allen, R.: Studies of the Social and Psychological Aspects of Verification, Inspection and International Assurance, Technical Report #4.1. Purdue University, Lafayette, IN (1968).

18. Dunn, M. H.: Trust and Political Agency. In: Gambetta, D. (ed.): Trust: Making and Breaking Cooperative Relations. Blackwell, New York (1988) 73-93.

19. Emerson, R. M.: Power-Dependence Relations. American Sociological Review, **27** (1962) 31-41.

20. Erikson, E. H.: Identity: Youth and Crisis. W. W. Norton, New York (1968).

21. Fishbein, M., Ajzen, I.: Belief, Attitude, Intention and Behavior: An Introduction to Theory and Research. Addison-Wesley, Reading, MA (1975).

22. Fox, A.: Beyond Contract: Work, Power and Trust Relations. Faber, London (1974).

23. French, J. R. P., Raven, B.: The Bases of Social Power. In: Cartwright, D. (ed.): Studies in Social Power. Institute for Social Research,, Ann Arbor, MI (1959) 150-167.

24. Gabarro, J. J.: The Development of Trust, Influence, and Expectations. In: Athos, A. G., Gabarro, J. J. (eds.): Interpersonal Behavior: Communication and Understanding in Relationships, Prentice-Hall, Englewood Cliffs, NJ (1978) 290-303.

25. Gaines, J. H.: Upward Communication in Industry: An Experiment. Human Relations. **33** (1980) 929-942.

26. Gambetta, D.: Can We Trust Trust? In: Gambetta, D. (ed.): Trust: Making and Breaking Cooperative Relations. Blackwell, New York (1988) 213-237.

27. Garfinkel, H.: A Conception of, and Experiments with, "Trust" as a Condition of Stable Concerted Actions. In: Harvey, O. J. (ed.): Motivation and Social Interaction, Ronald Press, New York (1963) 187-238.

28. Gellner, E.: Trust, Cohesion, and the Social Order. In: Gambetta, D. (ed.): Trust: Making and Breaking Cooperative Relations. Blackwell, New York (1988) 142-157.

29. Giffin, K.: The Contribution of Studies of Source Credibility to a Theory of Inter-personal Trust in the Communication Process. Psychological Bulletin **68** (1967) 104-120.
30. Golembiewski, R. T., McConkie, M.: The Centrality of Interpersonal Trust in Group Processes. In: Cooper, G. L. (ed.): Theories of Group Processes, John Wiley & Sons, London (1975) 131-185.
31. Good, D.: Individuals, Interpersonal Relations, and Trust. In: Gambetta, D. (ed.): Trust: Making and Breaking Cooperative Relations Blackwell, New York (1988) 31-48.
32. Granovetter, M.: Economic Action and Social Structure: The Problem of Embed-dedness, American Journal of Sociology **91** (1985) 481-510.
33. Hart, P. J., Saunders, C. S.: Themes of Power and Trust in EDI Relationships. In: DeGross, J. I., Bostrom, R. P., Robey, D. (eds.): Proceedings of the Fourteenth International Conference on Information Systems, Vol. 14. Orlando, FL (1993) 383.
34. Holmes, J. G.: Trust and the Appraisal Process in Close Relationships. In: Jones, W. H. , Perlman, D. (eds.): Advances in Personal Relationships, Vol. 2. Jessica Kingsley, London (1991) 57-104-
35. Hosmer, L. T.: Trust: The Connecting Link Between Organizational Theory and Philosophical Ethics. Academy of Management Review **20** (1995) 379-403.
36. Hoy, W. K., Kupersmith, W. J.: The Meaning and Measure of Faculty Trust. Educational and Psychological Research **5** (1985) 1-10.
37. Janis, I. L.: Victims of Groupthink. Houghton Mifflin, Boston (1972).
38. Jensen, M. C., Meckling, W. H.: Theory of the Firm: Managerial Behavior, Agency Costs and Ownership Structure. Journal of Financial Economics, **3** October (1976) 305-360.
39. Johnson-George, C., Swap, W. C.: Measurement of Specific Interpersonal Trust: Construction and Validation of a Scale to Assess Trust in a Specific Other. Journal of Personality and Social Psychology **43** (1982) 1306-1317.
40. Kaplan, A.: The Conduct of Inquiry. Chandler, New York (1964).
41. Kee, H. W., Knox, R. E.: Conceptual and Methodological Considerations in the Study of Trust and Suspicion, Journal of Conflict Resolution **14** (1970) 357-366.
42. Kegan, D. L., Rubenstein, A. H.: Trust, Effectiveness, and Organizational Devel-opment: A Field Study in R&D. Journal of Applied Behavioral Science **9** (1973) 495-513.
43. Kelley, H. H.: Common-Sense Psychology and Scientific Psychology. Annual Re-view of Psychology **43** (1992) 1-23.
44. Kipnis, D.: Trust and Technology. In: Kramer, R. M., Tyler, T. R. (eds.): Trust in Organizations: Frontiers of Theory and Research, Sage, Thousand Oaks, CA (1996) 39-50.
45. Krackhardt, D., Stern, R. N.: Informal Networks and Organizational Crises: Ex-perimental Simulation. Social Psychology Quarterly **51(2)** (1988) 123-140.
46. Kramer, R. M.: The Sinister Attribution Error: Paranoid Cognition and Collective Distrust in Organizations. Motivation and Emotion **18** (1994) 199-230.
47. Kuhn, T. S.: The Structure of Scientific Revolutions. University of Chicago Press, Chicago (1962).
48. Lewicki, R. J., Bunker, B. B.: Trust in Relationships: A Model of Trust Develop-ment and Decline. In: Bunker , B. B., Rubin, J. Z. (eds.): Conflict, Cooperation and Justice, Jossey-Bass, San Francisco (1995) 133-173.
49. Lewicki, R. J., McAllister, D. J., Bies, R. J.: Trust and Distrust: New Relation-ships and Realities. Academy of Management Review **23** (1998) 438-458.

50. Lewis, J. D., Weigert, A. J.: Trust as a Social Reality. Social Forces **63** (1985a) 967-985.
51. Lewis, J. D., Weigert, A. J.: Social Atomism, Holism, and Trust. The Sociological Quarterly **26** (1985b) 455-471.
52. Lindskold, S.: Trust Development, the GRIT Proposal, and the Effects of Conciliatory Acts on Conflict and Cooperation. Psychological Bulletin **8** July (1978) 772-793.
53. Loomis, J. L.: Communication, the Development of Trust, and Cooperative Behavior. Human Relations **12** (1959) 305-315.
54. Luhmann, N.: Trust and Power, John Wiley, New York (1979).
55. Luhmann, N.: Familiarity, Confidence, Trust: Problems and Alternatives. In: Gambetta, D. (ed.): Trust: Making and Breaking Cooperative Relations. Blackwell, New York (1988) 94-107.
56. Mancini, P.: Between Trust and Suspicion: How Political Journalists Solve the Dilemma. European Journal of Communication **8** (1993) 33-51.
57. Mayer, R. C., Davis, J. H., Schoorman, F. D.: An Integrative Model of Organizational Trust, Academy of Management Review **20** (1995) 709-734.
58. McAllister, D. J.: Affect- and Cognition-Based Trust as Foundations for Interpersonal Cooperation in Organizations. Academy of Management Journal **38** (1995) 24-59.
59. McGregor, D.: The Professional Manager. McGraw-Hill, New York (1967).
60. McKnight, D. H., Chervany, N. L.: What is Trust? A Conceptual Analysis and an Interdisciplinary Model, in Chung, Michael H. (ed.): Proceedings of the Americas Conference on Information Systems, August 10-13, 2000, Long Beach, California 827-833 (2000).
61. McKnight, D. H., Cummings, L. L. Chervany, N. L.: Trust Formation in New Organizational Relationships. MIS Research Center, Working Paper 96-01, Carlson School of Management, University of Minnesota (ftp://misrc.umn.edu/WorkingPapers/9601.pdf) (1996).
62. McKnight, D. H., Cummings, L. L., Chervany, N. L.: Initial Trust Formation in New Organizational Relationships. Academy of Management Review **23** (1998) 473-490.
63. McKnight, D. H., Sitkin, S. B., Chervany, N. L.: How Familiarity and Social Categorization Affect the Dimensionality of Trusting Beliefs: Two Tests. Unpublished working paper, Florida State University (2001).
64. McLain, D. L., Hackman, B. K.: Trust and Risk Taking in Organizations. Unpublished working paper, Virginia State University (1995).
65. Mellinger, G. D.: Interpersonal Trust as a Factor in Communication. Journal of Abnormal and Social Psychology **52** 304-309 (1956).
66. Mishra, A. K.: Organizational Responses to Crisis: The Centrality of Trust. In: Kramer, R. M., Tyler, T. R. (eds.): Trust in Organizations: Frontiers of Theory and Research. Sage, Thousand Oaks, CA (1996) 261-287.
67. Mitchell, S.: The Official Guide to American Attitudes: Who Thinks What About the Issues that Shape Our Lives. New Strategist Publications, Ithaca, NY (1996).
68. Nelson, K. M., Cooprider, J. G.: The Contribution of Shared Knowledge to IS Group Performance. MIS Quarterly **20** (1996) 409-434.
69. O'Reilly, C. A.: The Intentional Distortion of Information in Organizational Communication: A Laboratory and Field Investigation. Human Relations **31** 173-193 (1978).
70. Osigweh, C.: Concept Fallibility in Organizational Science, Academy of Management Review **14** (1989) 579-594.

71. Ouchi, W. G.: Theory Z: How American Business Can Meet the Japanese Challenge. Addison-Wesley, Reading, MA (1981).
72. Rempel, J. K., Holmes, J. G., Zanna, M. P.: Trust in Close Relationships, Journal of Personality and Social Psychology **49** (1985) 95-112.
73. Riker, W. H.: The Nature of Trust. In: Tedeschi, J. T. (ed.): Perspectives on Social Power. Aldine Publishing Company, Chicago (1971) 63-81.
74. Ring, P. S., Van de Ven, A. H.: Developmental Processes of Cooperative Interorganizational Relationships. Academy of Management Review **19** (1994) 90-118.
75. Roberts, K. H., O'Reilly, C. A.: Measuring Organizational Communication. Journal of Applied Psychology, **59** (1974) 321-326.
76. Rosenberg, M.: Occupations and Values. Free Press, Glencoe, IL (1957).
77. Rotter, J. B.: A New Scale for the Measurement of Interpersonal Trust. Journal of Personality **35** (1967) 651-665.
78. Rotter, J. B.: Generalized Expectancies for Interpersonal Trust. American Psychologist **26** (1971) 443-452.
79. Rotter, J. B.: Interpersonal Trust, Trustworthiness, and Gullibility. American Psychologist, **35** (1980) 1-7.
80. Rousseau, D. M., Sitkin, S. B., Burt, R. S., Camerer, C.: Not so Different after All: A Cross-Discipline View of Trust. Academy of Management Review **23** (1998) 393-404.
81. Rubin, Z.: Liking and Loving: An Invitation to Social Psychology. Holt, Rinehart and Winston, New York (1973).
82. Sagasti, F. R., Mitroff, I. I.: Operations Research from the Viewpoint of General Systems Theory. OMEGA, 1 (1973) 695-709.
83. Sato, K: Trust and Group Size in a Social Dilemma. Japanese Psychological Research, **30** 88-93 (1988).
84. Scanzoni, J.: Social Exchange and Behavioral Interdependence. In: Burgess, R. L., Huston, T. L. (eds.): Social Exchange in Developing Relationships, Academic Press, New York (1979) 61-98.
85. Schwab, D. P.: Construct Validity in Organizational Behavior. In: Staw, B. M., Cummings, L. L. (eds.): Research in Organizational Behavior, Vol. 2. JAI Press, Greenwich, CN (1980) 3-43.
86. Shapiro, S P.: The Social Control of Impersonal Trust. American Journal of Sociology **93** (1987) 623-658.
87. Sheppard, B. H., Tuchinsky, M.: Interfirm Relationships: A Grammar of Pairs. In: Staw, B. M. and Cummings, L. L. (eds.): Research in Organizational Behavior, Vol. 18. JAI Press, Greenwich, CN (1996) 331-373.
88. Shapiro, D. L., Sheppard, B. H., Cheraskin, L.: Business on a Handshake. Negotiation Journal, **8** (1992) 365-377.
89. Sitkin, S. B. , Bies, R. J. (eds.): The Legalistic Organization. Sage, Thousand Oaks, CA (1994).
90. Sitkin, S. B., Roth, N. L.: Explaining the Limited Effectiveness of Legalistic "Remedies" for Trust / Distrust. Organization Science 4 (1993) 367-392.
91. Solomon, L.: The Influence of Some Types of Power Relationships and Game Strategies Upon the Development of Interpersonal Trust. Journal of Abnormal and Social Psychology **61** (1960) 223-230.
92. Starbuck, W. H., Milliken, F. J.: Challenger: Fine-tuning the Odds until Something Breaks. Journal of Management Studies **25** (1988) 319-340.
93. Taylor, R. G.: The Role of Trust in Labor-Management Relations. Organization Development Journal summer, (1989) 85-89.

94. Tiryakian, E. A.: Typologies. In: Sills, D. L. (ed.): International Encyclopedia of the Social Science, Vol. 16. The Macmillan Company & The Free Press (1968) 177-186.

95. Van de Ven, A. H.: Nothing is Quite so Practical as a Good Theory. Academy of Management Review **14** (1989) 486-489.

96. Van Dyne, L., Vandewalle, D., Kostova, T., Latham, M. E., Cummings, L. L.: Collectivism, Propensity to Trust and Self-Esteem as Predictors of Organizational Citizenship in a Non-Work Setting. Journal of Organizational Behavior **21** (2000) 3-23.

97. Walton, R. E.: Social and Psychological Aspects of Verification, Inspection, and International Assurance, Technical Report 1. Purdue University, Lafayette (1968).

98. Webb, E. J.: Trust and Crisis. In: Kramer, R. M., Tyler, T. R. (eds.): Trust in Organizations: Frontiers of Theory and Research. Sage, Thousand Oaks, CA (1996) 288-301.

99. Williamson, O. E.: Markets and Hierarchies: Analysis and Antitrust Implications. Free Press, New York (1975).

100. Williamson, O. E.: Calculativeness, Trust, and Economic Organization. Journal of Law and Economics **34** (1993) 453-502.

101. Worchel, P.: Trust and Distrust. In: Austin, W. G., Worchel, S. (eds.): The Social Psychology of Intergroup Relations. Wadsworth, Belmont, CA (1979) 174-187.

102. Wrightsman, L. S.: Interpersonal Trust and Attitudes toward Human Nature. In: Robinson, J. P., Shaver, P. R., Wrightsman, L. S. (eds.): Measures of Personality and Social Psychological Attitudes: Vol 1. Measures of Social Psychological Attitudes, Academic Press, San Diego, CA (1991) 373-412.

103. Yamagishi, T., Yamagishi, M.: Trust and Commitment in the United States and Japan. Motivation and Emotion **18** (1994) 129-166.

104. Zand, D. E.: Trust and Managerial Problem Solving. Administrative Science Quarterly **17** (1972) 229-239.

105. Zucker, L. G.: Production of Trust: Institutional Sources of Economic Structure, 1840-1920. In Staw, B. M., Cummings, L. L. (eds.): Research in Organizational Behavior, Vol. 6. JAI Press, Greenwich, CN (1986) 53-111.

The Socio-cognitive Dynamics of Trust: Does Trust Create Trust?

Rino Falcone[1] and Cristiano Castelfranchi[2]

[1] IP-CNR
falcone@ip.rm.cnr.it
http://www.ip.rm.cnr.it/iamci/falcone.htm
[2] University of Siena
castel@ip.rm.cnr.it

Abstract. We[1] will examine in this paper three crucial aspects of trust dynamics:

a) *How A's trusting B and relying on it in situation Ω can actually (objectively) influnce B's trustworthiness within Ω* . Either trust is a self-fulfilling prophecy that modifies the probability of the predicted event; or it is a self-defeating strategy by negatively influencing the events. And also how A can be aware of and take into account the effect of its own decision in the very moment of that decision.

b) *How trust creates a reciprocal trust, and distrust elicits distrust*; but also vice versa: how A's trust in B could induce lack of trust or distrust in B towards A, while A's diffidence can make B more trustful in A. And also how A can be aware of and take into account this effect of its own decision in the very moment of that decision.

c) *How diffuse trust diffuses trust (trust atmosphere)*, that is how A's trusting B can influence C trusting B or D, and so on.

Those phenomena are very crucial in human societies (market, groups, states), however we claim that they are also very fundamental in computer mediated organizations, interactions (like Electronic Commerce), cooperation (Computer Supported Cooperative Work), etc. and even in Multi-Agent Systems with autonomous agents.

1 Premise

There is today a diffuse agreement about the fact that one of the major problems for the success of computer supported society, smart physical environment, virtual reality, virtual organization, computer mediated interaction (like EC or CSCW), etc. is *trust*: people's trust in the computational infrastructure; people's trust in potential partners, information sources, data, mediating agents,

---

[1] This paper has been partially developed within the European Project ALFEBIITE (A Logical Framework for Ethical Behaviour between Infohabitants in the Information Trading Economy of the Universal Information Ecosystem): IST-1999-10298; and in part by the TICCA Project (joint research venture between the Italian National Research Council -CNR- and Provincia Autonoma di Trento).

R. Falcone, M. Singh, and Y.-H. Tan (Eds.): Trust in Cyber-societies, LNAI 2246, pp. 55–72, 2001.

personal assistants; and agents' trust in other agents and processes. Security measures are not enough, interactivity and knowledgability are not enough; the problem is how to build in users and agents trust and how to maintain it.

Building trust is not just a matter of protocols, architectures, mind-design, clear rules and constraints, controls and guaranties. Trust in part is a socially emergent phenomenon; it is a mental stuff but in socially situated agents and based on social context. In particular, trust is a very dynamic phenomenon; not only because it evolves in time and has a history, that is A"s trust in B depends on A's previous experience and learning with B itself or with other (similar) entities [1, 2, 3], but because trust is not simply an external observer's prediction or expectation about a matter of fact. In one and the same situation *trust is influenced by trust* in several rather complex ways. We will examine in this paper three crucial aspects of trust dynamics:

a) *How A" s trusting B and relying on it in situation Ω can actually (objectively) influence B's trustworthiness in Ω* . Either trust is a self-fulfilling prophecy that modifies the probability of the predicted event; or it is a self-defeating strategy by negatively influencing the events. And also how A can be aware of and take into account the effect of its own decision in the very moment of that decision.

b) *How trust creates a reciprocal trust, and distrust elicits distrust*; but also vice versa: how A"s trust in B could induce lack of trust or distrust in B towards A, while A"s diffidence can make B more trustful in A. And also how A can be aware of and take into account this effect of its own decision in the very moment of that decision.

c) *How diffuse trust diffuses trust (trust atmosphere)*, that is how A"s trusting B can influence C trusting B or D, and so on. Usually, this is a macro-level phenomenon and the individual agent does not calculate it.

As we have argued in previous works [4, 5, 6] trust and reliance/delegation are strictly connected phenomena: trust could be considered as the set of mental components a delegation action is based on. So, also in this analysis of the dynamic of trust, we have to consider the role of the various possible kinds of delegation (weak, mild and strong) [7].

2 How Trust and Delegation Change the Trustee's Trustworthiness

In this section we are going to analyze how a delegation action (corresponding to a decision making based on trust in a specific situational context) could change the trustworthiness of the delegated agent (*delegee*).

Let us introduce the needed formal costructs. We define $Act = \{\alpha_1, ..., \alpha_n\}$ be a finite set of *actions*, and $Agt = \{X, Y, A, B, ...\}$ a finite set of *agents*. Each agent has an action repertoire, a plan library [17, 21], resources, goals (in general by "goal" we mean a partial situation (a subset) of the "world state"), beliefs, motives, etc..

We assume that *to delegate an action necessarily implies delegating some result of that action* (i.e. expecting some results from X's action and relaying on it for obtaining those results). Conversely, *to delegate a goal state always implies the delegation of at least one action (possibly unknown to Y) that produces such a goal state as result* (even when Y asks X to solve a problem, to bring it about that g, without knowing or specifying the action, Y necessarily presupposes that X should and will do some action and relies on this).

Thus, we consider the action/goal pair $\tau = (\alpha, g)$ as the real object of delegation, and we will call it 'task'. Then by means of τ, we will refer to the action (α), to its resulting world state (g), or to both.

Given an agent X and a situational context Ω (a set of propositions describing a state of the world), we define as trustworthiness of X about τ in Ω (called *trustworthiness* $(X\,\tau\,\Omega)$), the objective probability that X will succesfully execute the task τ in context Ω. This objective probability is in terms of our model [6] computed on the basis of some more elementary components:

- a degree of ability (*DoA*, ranging between O and 1, indicating the level of X's ability about the task τ); we can say that it could be measured as the number of X's successes (s) on the number of X's attempts (a): s/a, when a goes to ∞; and
- a degree of willingness (*DoW*, ranging between O and 1, indicating the level of X's intentionality/persistance about the task τ); we can say that it could be measured as the number of X's (succesfully or unsuccessfully) performances (p) of that given task on the number of times X declares to have the intention (i) to perform that task: p/i, when i goes to ∞; we are considering that an agent declares its intention each time it has got one.

So, in this model we have that:

$$trustworthiness\,(B\,\tau\,\Omega) = F(DoA_{B\,\tau\,\Omega}, DoW_{B\,\tau\,\Omega}) \tag{1}$$

where F is in general a function that preserves monotonicity: for the purpose of this work it is not relevant to analyze the various possible models of the function F. We have considered this probability as *objective* because we hypothesize that it measures the real value of the X's trustworthiness; for example, if *trustworthiness* $(X\,\tau\,\Omega) = 0.80$, we suppose that in a context Ω, 80% of times X tries and succedes in executing τ.

As the reader can see we have not considered the opportunity dimension: the external (to the agent) conditions allowing or inhibiting the realization of the task.

2.1   The Case of Weak Delegation

We call *weak delegation* the reliance simply based on exploitation for the achievement of the task. In it there is no agreement, no request or even (intended) influence: A is just exploiting in its plan a fully autonomous action of B.

The expression $W\text{-}Delegates\,(A\,B\,\tau)$ represents the following *necessary* mental ingredients:

a) The achievement of τ (the execution of α and its result g) is a *goal* of A.
b) A believes that there exists another agent B that has the *power of* [8] achieving τ .
c) A believes that B will achieve τ in time and by itself (without any A's intervention).
c-bis) A believes that B *intends* (in the case that B is a cognitive agent) to achieve τ in time and by itself, and that will do this in time and without any A's intervention.
d) A *prefers*[2] to achieve τ through B.
e) The achievement of τ through B is the choice (goal) of A.
f) A has the goal (*relativized* [11] to (e)) of not achieving τ by itself.

We consider (a, b, c, and d) what the agent A views as a "*Potential for relying on*" the agent B, its *trust* in B; and (e and f) what A views as the "*Decision to rely on*" B. We consider "Potential for relying on" and "Decision to rely on" as two constructs temporally and logically related to each other.

We hypothesize that in weak delegation (as in any delegation) there is a decision making based on trust and in particular there are two specific beliefs of A:

belief1 : if B makes the action then B has a successfull performance;
belief2 : B intend to do the action;

and A accepts these beliefs with a certain degree of credibility:

- the subjective (following A) degree of delegee's abilities $S_A DoA_{B\,\tau\,\Omega}$, and
- the subjective (following A) degree of delegee's willingness $S_A DoW_{B\,\tau\,\Omega}$.

So we can say that the trustworthiness of B by A $(DoT_{A\,B\,\tau})$ is:

$$DoT_{A\,B\,\tau} = F'(S_A DoA_{B\,\tau\,\Omega},\ S_A DoW_{B\,\tau\,\Omega}) \qquad (2)$$

where F' is function that preserves monotonicity; it is in general different from F. For sake of semplicity we assume that:

$$DoT_{A\,B\,\tau} \equiv trustworthiness\,(B\,\tau\,\Omega) \qquad (3)$$

in other words, A has a perfect perception of the B's trustworthiness.

The interesting case in weak delegation is when:

$$Bel(A\,\neg\,Bel(B\ W\text{-}Delegates\,(A\,B\,\tau)))\,\cap\,Bel(B\ W\text{-}Delegates\,(A\,B\,\tau))^3 \qquad (4)$$

---

[2] This means, either relative to the achievement of τ or relative to a broader goal g' that includes the achievement of τ, A believes to be dependent on B [9,10]
[3] Other possible alternative hypothesis are:
$\neg\,Bel(A\ Bel(B\ W\text{-}Delegates\,(A\,B\,\tau)))\,\cap\,Bel(B\ W\text{-}Delegates\,(A\,B\,\tau))$
or
$Bel(A\ Bel(B\,\neg\ W\text{-}Delegates\,(A\,B\,\tau)))\,\cap\,Bel(B\ W\text{-}Delegates\,(A\,B\,\tau))$

in words, there is a weak delegation by A on B and B is awareness of it (while A believes that B is not awareness). The first belief is always true in weak delegation, while the second one is necessary in the case we are going to consider. In fact, if

$$Bel(B \text{ } W\text{-}Delegates \text{ } (A \text{ } B \text{ } \tau)) \qquad (5)$$

this belief could change the B's trustworthiness, either because B will adopt A's goal and accept such a reliance/exploitation, or because B will react to that. After the action of delegation we in fact have a new situation Ω' (we assume that the delegation is the only event that has an influence on the trustworthiness) and we can have two possible results:

i) the new trustworthiness of B as for τ is greater than the previous one; at least one of the two possible elementary components is increased: DoA, DoW; so we can write:

$$\Delta \text{ } trustworthiness \text{ } (B \text{ } \tau) =$$
$$= F(DoA_{B \text{ } \tau \text{ } \Omega'}, DoW_{B \text{ } \tau \text{ } \Omega'}) - F(DoA_{B \text{ } \tau \text{ } \Omega}, DoW_{B \text{ } \tau \text{ } \Omega}) > 0 \qquad (6)$$

ii) the new reliability of B as for τ is less than the previous one.

$$\Delta \text{ } trustworthiness \text{ } (B \text{ } \tau) < 0. \qquad (7)$$

In case (i) B has adopted A's goal, i.e. it is doing τ <u>also</u> in order to let/make A achieve its goal g. Such adoption of A's goal can be for several possible motives, from instrumental and selfish, to pro-social. There are several ways of changing the degree of the two components:

- the degree of ability (DoA) can increase because B could use additional tools, new consulting agents, and so on;
- the degree of willingness (DoW) can increase because B could have more attention, intention, and so on.

In case (ii) B on the contrary reacts in a negative (for A) way to the discover of A's reliance and exploitation; for some reason B is now less willing or less capable in doing τ. In fact in case (ii) too, the reliability components can be independently affected:

- the degree of ability (DoA) can decrease because B could be upset about the A's exploitation and even if B does not want reduce its ability, it could result compromised;
- the degree of willingness (DoW) can decrease because B will have less intention, attention, etc.

Notice that in this case the change of the B's reliability is not known by A. So, even if A has a good perception of previous B's trustworthiness (that is our hypothesis), in this new situation -with weak delegation- A can have an under or over estimation of B's trustworthiness. In other terms, after the weak delegation (and if there is a change of B's trustworthiness following it) we have:

$$DoT_{A \text{ } B \text{ } \tau} \neq trustworthiness \text{ } (B \text{ } \tau) \qquad (8)$$

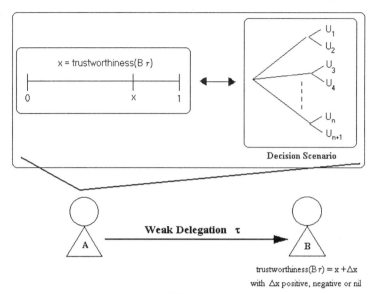

Fig. 1.

In Fig.1 is resumed how weak delegation can influence the delegee's trust-worthiness: agent A has both a belief about the B's trustworthiness and a hy-pothetical scenario of the utilities (in the case of success or failure) of all the possible choices it can do (to delegate to B or to C, etc., or do not delegate and doing by itself or doing nothing). On this basis it makes a weak delegation and may be it changes the B's trustworthiness.

2.2   The Case of Strong Delegation

We call *strong delegation*, that based on explicit agreement, i.e. on the achieve-ment by A of the task through an agreement with B.

The expression *S-Delegates* $(A \, B \, \tau)$ represents the following *necessary* mental ingredients:

a") The achievement of τ is a *goal* of A.

b") A believes that there exists another agent B that has the *power of* achieving τ.

c") A does not believe that B will achieve τ by itself (without any A's inter-vention).

d") A believes that if A realizes an action α' there will be this result: B will intend τ as the consequence of the fact that B adopts the A's goal that B would intend τ (in other words, B will be socially committed with A).

e") A *prefers* to achieve τ through B.

f") A intends to do α' relativized to (d").

g") The achievement of τ through B is the goal of A.

h") A has the goal (*relativized* to (g')) of not achieving τ by itself.

We consider (a", b", c", d" and e") what the agent A views as a *"Potential for relying on"* the agent B; and (f", g" and h") what A views as the *"Decision to rely on"* B.

In strong delegation we have:

$$MBel\,(A\,B\,S\text{-}Delegates\,(A\,B\,\tau)) \qquad (9)$$

i.e. there is a mutual belief of A and B about the strong delegation and about the reciprocal awareness of it.

Like in the weak delegation this belief could change the B's trustworthiness, and also in this case we can have two possible results:

i) the new trustworthiness of B as for τ is greater than the previous one; so we can write:

$$\Delta\,trustworthiness\,(B\,\tau) =$$
$$= F(DoA_{B\,\tau\,\Omega'}, DoW_{B\,\tau\,\Omega'}) - F(DoA_{B\tau\Omega}, DoW_{B\tau\Omega}) > 0 \qquad (10)$$

ii) the new trustworthiness of B on τ is less than the previous one.

$$\Delta\,trustworthiness\,(B\,\tau) < 0. \qquad (11)$$

Why does B's trustworthiness increase or decrease? What are the reasons? In general a strong delegation increases the trustworthiness of the delegee because of its *commitment* [12]. This is in fact one of the motive why agents use strong delegation. But it is also possible that the delegee loses motivations when it has to do something not spontaneously but by a contract or by a role.

The important difference with the previous case is that now A knows that B will have some possible reactions to the delegation and consequently A is expecting a new B's trustworthiness. So a strong delegation can have influence on the B's trustworthiness and it is also possible that A is able to know this new B's trustworthiness (our hypothesis) so that A can update its DoT (see Fig.2):

$$DoT_{A\,B\,\tau} = trustworthiness\,(B\,\tau) \qquad (12)$$

Another interesting case, that regards again strong delegation, is when there is the decision to rely upon B but with diffidence and without any certainty that B will be able to achieve τ. We are interested in the case in which B realizes that diffidence. We have:

$$S\text{-}Delegates\,(A\,B\,\tau) \qquad (13)$$

with $DoT_{A\,B\,\tau} < \sigma$; where σ is a "reasonable threshold" to delegate the task τ and $Bel\ \ (B\,(DoT_{A\,B\,\tau} < \sigma))$.

Such a diffidence could be implicit or explicit in the delegation: it is not important. Neither it is important, in this specific analysis, if

$$DoT_{A\,B\,\tau} \neq trustworthiness\,(B\,\tau) \qquad (14)$$

or

$$DoT_{A\,B\,\tau} = trustworthiness\,(B\,\tau) \qquad (15)$$

in other words, if this diffidence is objectively justified or not. This distrust could, in fact, produce a change (either positive or negative) in B's trustworthiness:

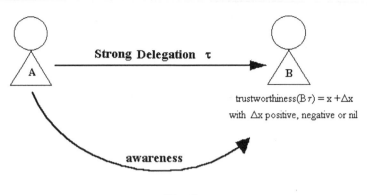

Fig. 2.

- B could be disturbed by such a bad evaluation and have a worst performance (we will not consider here, but in general in these cases of diffidence there is always some additional actions by the delegating agent: more control, some parallel additional delegation, and so on);
- B even if disturbed by the diffidence could have a pridefully reaction and produce a better performance.

Resuming, B's trustworthiness could be changed by both A's trust or distrust.

2.3   Anticipated Effects of the Reliability Changes

This is the case in which the delegating agent A takes into account the possible effects of its strong delegation on the B's trustworthiness before it performs the delegation action itself: in this way A changes the $DoT_{AB\tau}$ before the delegation (Fig.3).

We could have two main interesting subcases:

i) the new degree of trust $(DoT'_{AB\tau})$ is greater than the old one $(DoT_{AB\tau})$:

$$DoT'_{AB\tau} > DoT_{AB\tau}; \qquad (16)$$

ii) the new degree of trust is less than the old one:

$$DoT'_{AB\tau} < DoT_{AB\tau}. \qquad (17)$$

In table 1 are considered all the possible decisions of A.

We have called σ the minimum threshold to delegate. In other words, before to perform a delegation action and just for delegating, an agent A could evaluate the influence (positive or negative) that its delegation will have on the B's trustworthiness. Only after this evaluation A makes the decision.

Another interesting way for increasing this trustworthiness it is through the self-confidence dimension, that we did not explicitly mention since it is part of

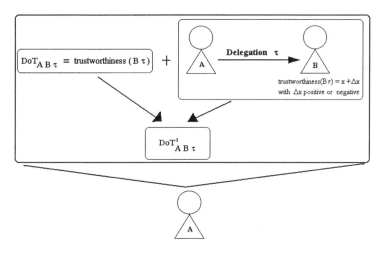

Fig. 3.

the ability dimension. In fact, at least in human agents the ability to do α is not only based on skills (an action repertoire) or on knowing how (library of recipes, etc.), it also requires self-confidence that means the subjective awareness of have those skills and expertise, plus a general good evaluation (and feeling) of its own capability of success. Now the problem is that self-confidence is socially influenced, i.e. my confidence and trust in you can increase your self-confidence. So, I could strategically rely on you (letting you know that I'm relying on you)

Table 1.

	DoT $> \sigma$ DoT $-\sigma = \epsilon' > 0$ (*A* would delegate *B* before evaluating the effects of delegation itself)	DoT $< \sigma$ DoT $-\sigma = \epsilon' < 0$ (*A* would not delegate *B* before evaluating the effects of delegation itself)
DoT' - DoT $= \epsilon > 0$ (*A* thinks that the delegation would increase *B*'s trustworthiness)	DoT' $-\sigma = \epsilon + \epsilon' > 0$ **Decision to delegate**	DoT' $-\sigma = \epsilon - \epsilon' > 0$ $(\epsilon > \epsilon')$ **Decision to delegate** DoT' $-\sigma = \epsilon - \epsilon' < 0$ $(\epsilon < \epsilon')$ **Decision not to delegate**
DoT' - DoT $= \epsilon < 0$ (*A* thinks that the delegation would decrease *B*'s trustworthiness)	DoT' $-\sigma = \epsilon - \epsilon' > 0$ $(\epsilon' > \epsilon)$ **Decision to delegate** DoT' $-\sigma = \epsilon - \epsilon' < 0$ $(\epsilon' < \epsilon)$ **Decision not to delegate**	DoT' $-\sigma = -\epsilon - \epsilon' < 0$ **Decision not to delegate**

in order to increase your self-confidence and then my trust in you as for your ability and trustworthiness to do.

3   The Dynamics of Reciprocal Trust and Distrust

The act of trusting somebody (i.e. the reliance) can be also an implicitly communicative act. This is especially true when the delegation is *strong* (when it implies and relies on the understanding and agreement of the delegee), and when it is part of a bilateral and possibly reciprocal relation of delegation-help, like in social exchange. In fact in *social exchange* A's adoption of B's goal is *conditional* to B's adoption of A's goal. A's adoption is based on A's trust in B, and vice versa. Thus, A's trusting B for delegating to B a task τ is in some sense conditional on B's trusting A for delegating to A a task τ', A has also to trust (believe) that B will trust it; and vice versa. There is a recursive embedding of trust attitudes. Not only this, but the measure of A's trusting B depends on, varies with the decision and the act of B's trusting A (and vice versa). *The act of trusting can have among its effects that of determining or increasing B's trusting A.* Thus A may be aware of this effect and may plan to achieve it through its act of trusting. In this case A must plan for B understanding its decision/act of trusting B. But, why wants A communicate to B about its decision and action of relying on B? In order to induce some (more) trust in B. Thus the higher goal of that communication goal in A's plan is to induce B to believe that "B can trust A since A trusts B". And this is eventually in order (to higher goal) to induce B to trust A. As claimed in sociology [13] there is in social relations the necessity of actively promoting trust and ""the concession of trust" - that generates precisely that behaviour that seems to be its logical presupposition- is part of a strategy for structuring social exchange. In sum, usually there is a circular relation, and more precisely a positive feedback, between trust in reciprocal delegation-adoption relations (from commerce to friendship). That -in cognitive terms- means that the (communicative) act of trusting and eventually delegating impacts on the beliefs of the other ("trust" in strict sense) that are the bases of the ""reliance"" attitude and decision producing the external act of delegating (Fig.4).

Analogously there is a positive feedback relations between distrust: as usually trust induces trust in the other, so usually distrust increments distrust. Which precisely is the mechanism of trust producing trust in exchange?[4] In my trust about your willingness to exchange (help & delegation) is included my trust that you trust me (and then delegate and then accept to help): if you do not trust me, you will do not delegate me and then you will not adopt my goal. And vice

---

[4] More in general, in reciprocal relationship, trust elicits trust. This is also typical in friendship. If he trusts me (as for keeping secrets, don't cheating, don't laughing) he cannot have aggressive motives against me (he will expose himself to retaliation; he should feel antipathy, but antipathy does not support confidence). So he must be benevolent, keep secrets, don't cheat, don't laugh: this is my trust in reciprocating confidences.

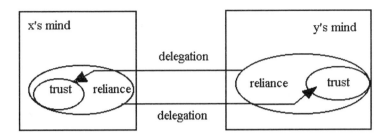

Fig. 4.

versa. Now my trusting you (your believing that I trust you) may increase your trust in me as for delegating for exchange: since I delegate you, (conditionally, by exchange) you can believe that I am willing to adopt your goal, so you can delegate me. There are also other means to change B's willingness to trust us a nd delegate to us (and then for example exchange with us). Suppose for example that I am not sure about B trusting me because of some bad reputation or some prejudice about me. I can change B's opinion about me (ex. through some recommendation letter, or showing my good behaviour in other interactions) in order to improve B's trust in me and then B's willingness, and then my own trust in delegating to it.

An interesting concept is the so-called *reciprocal trust* that is not simply *bilateral trust*. It is not sufficient that X trusts Y and Y trusts X at the same time. For example, X relies on Y for stopping the bus and Y relies on X for stopping the bus, there is bilateral (unaware) trust and delegation; nobody stops the bus and both fail!

To have *reciprocal trust*, mutual understanding and communication (at least implicit) are needed: X has the goal that Y knows that it (will) trust Y, in order Y trusts it, and that it trust Y only if Y trusts it.

Exchange is in fact characterised by *reciprocal conditional trust*: how the act of trusting can increase B's trust and then my own trust which should be presupposed by my act of trusting. However no paradox or irrationality is there, since my prediction of the effect anticipates my act and justifies it.

Given our agents A and B and two possible tasks: τ and τ', we suppose that:

$$DoT_{AB\tau} < \sigma_1.^5 \tag{18}$$

The value of the $DoT_{AB\tau}$ changes on the basis of its components' variation. One of the ways to change the elementary components of $DoT_{AB\tau}$ is when the trustee (B in our case) communicates (explicitly or implicitly) its own trust in the trustier (A) as for another possible task (τ') for example delegating the task τ' to A(relying upon A as for the task τ'). In our terms:

$$S\text{-}Delegation\,(B\,A\,\tau') \tag{19}$$

---

[5] where σ_1 is the A's reasonable threshold for delegating to B the task τ.

that always implies

$$MutualBel(A\ B\ S\text{-}Delegation(B\ A\ \tau')) \qquad (20)$$

In fact, this belief has various possible consequences in the A's mind:

i) there exists a dependence relationship between A and B and in particular the B 's achievement of the task τ' depends by A. Even if it is important to analyze the nature of the B's delegation as for τ' [6] , in general A has the awareness to have any power on B (and that B believes this).

ii) if this delegation is spontaneous and in particular it is a special kind of delegation (for example it is a display of esteem) and A has awareness of this, i.e. $Bel\ (A\ DoT_{B\,A\,\tau'} > \sigma_2)$[7] , in general an abstract benevolence could arise in A as for B.

iii) A could imply from point (i) the B's unharmfullness (if the delegation nature permits it).

iv) Trusting as a sign of goodwill and trustwortiness: Agents with bad intentions are frequently diffident towards the others; for example, in contracts they specify and check everything. Since they have non-benevolent intentions they ascribe similar attitudes to the others. This is why we believe that malicious agent are usually diffident and that (a risky abduction) suspicious agents are malicious. On such a basis we also feel more trustfull towards a non-diffident, trusting agent: this is a sign for us that it is goodwilling, non-malicious.

Each of the previous points allows the possibility that A delegates to B τ: Going to analyze the specific components of the degree of trust we can say that:

- the point (i) could increase both $S_A\ DoW_{B\,\tau}$ and $S_A\ DoA_{B\,\tau}$;
- the point (ii) and (iii) decrease the value of σ_1 and may be increase both $S_A DoW_{B\,\tau}$ and $S_A DoA_{B\,\tau}$.

In other words, after the B's delegation we can have two new parameters: $DoT'_{A\,B\,\tau}$ and σ_1' instead of $DoT_{A\,B\,\tau}$ and σ_1 and it is possible that:
$$DoT'_{A\,B\,\tau} > \sigma_1' .$$

Another interesting case is when a B's diffidence in A it is believed by A itself:

$$Bel(A\,DoT_{B\,A\,\tau'} < \sigma_2) \qquad (21)$$

and for this

$$Bel(A\ \neg S\text{-}Delegates\,(B\ A\ \tau')) \qquad (22)$$

Also in this case various possible consequences in the A's mind are possible:

i) the fact that B has decided to not depend by A could imply the B's willingness to avoid possible retaliation by A itself; so that A could imply the possible B's harmfullness.

---

[6] for example if there is already an agreement between A and B about τ' with reciprocal commitments and possible sanctions in the case in which there could be some uncorrect behaviour.

[7] where σ_2 is the B's reasonable threshold for delegating to A the task τ'.

Table 2.

	$DoT_{AB\tau} > \sigma_1$	$DoT_{AB\tau} < \sigma_1$
Bel $(A$ S-Delegates$(B\,A\,\tau'))$ and/or Bel $(A$ DoT$_{B\,A\,\tau'} > \sigma_2)$	Increases the A's degree of trust in B as for τ	Increases the $DoT_{AB\tau}$ but $\neg$ Delegates $(A\,B\,\tau)$ Increases the $DoT_{AB\tau}$ and Delegates $(A\,B\,\tau)$
Bel $(A\,\neg$Delegation$(B\,A\,\tau'))$ and/or Bel $(A$ DoT$_{B\,A\,\tau'} < \sigma_2)$	Decreases the $DoT_{AB\tau}$ but Delegates (A B τ) Decreases the $DoT_{AB\tau}$ and $\neg$ Delegates $(A\,B\,\tau)$	Decreases the A's degree of trust in B as for τ

ii) a B's expression of not estimation of A (and the fact that Bel $(A\,DoT_{B\,A\,\tau'} < \sigma_2)$, in general has the consequence that an abstract malevolence could arise in A as for B.

In table 2 we resume all the various possibilities.

4   The Diffusion of Trust: Authority, Example and Contagion

We examine here the point:
how diffuse trust diffuses trust (trust atmosphere), that is how A's trusting in B can influence C trusting B or D, and so on. Usually this is a macro-level phenomenon and the individual agent does not calculate it.

Let us consider two prototypical cases, the two micro-constituents of the macro process:

i) Since A trusts B, also C trusts B
ii) Since A trusts B, (by analogy) C trusts D.

It should be clear the potential multiplicative effects of those mechanisms/rules: (i) would lead to a trust network like Fig.5(A), while (ii) would lead to a structure like Fig.5(B)

4.1   Since B Trusts C, also A Trust C

There are at least two mechanisms for this form of spreading of trust.
A's authority (pseudo-transitivity)
We start from the following situation:

$$Bel(A\,DoT_{B\,C\,\tau} > \sigma_2) \qquad (23)$$

that is:

$$Bel(A\,(S_B\,DoO_{C\,\tau} \cdot S_B\,DoA_{C\,\tau} \cdot S_B\,DoW_{C\,\tau}) > \sigma_2) \qquad (24)$$

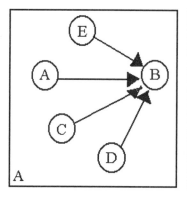 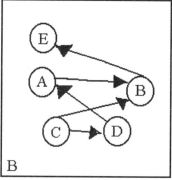

Fig. 5.

In words: agent A believes that the B's degree of trust in C on the task τ is greater than a reasonable threshold (following B). And more specifically A believes that the composition of the various components of the B's degree of trust is greater than B's trust threshold.

Given this A's belief, an interesting question is: what about the $DoT_{AC\tau}$? Is there a sort of trust transitivity? If it is so, what is its actual nature? Let us consider the case in which the only knowledge that A could have about C and about the possible C's performance of τ is given by B. We hypothesize that:

$$DoT_{AC\tau} = Inf_A\left(B\,C\,e(\tau)\right) \cdot DoT_{A\,B\,e(\tau)} \cdot DoT_{BC\tau} \qquad (25)$$

where:

$DoT_{A\,B\,e(\tau)}$ is the A's degree of trust in B about a new task $e(\tau)$ that is the task of generally evaluate competences, opportunities, etc. on τ. And $Inf_A\left(B\,C\,e(\tau)\right)$ represents the potential B's influence in the task $e(\tau)$ when τ is performed by C (following A's opinion).

Notice that this pseudo-transitivity depends on subtle cognitive conditions. In fact it is not enough that A trusts B for adopting its trusting attitude towards C. Suppose for example that A trusts B as a medical doctor but considers B a very impractical person as for business, and A knows that B trusts C as stock-market broker agent; A has no reasons for trusting C. On the contrary, if following A B is a good doctor and he trusts C as a good nurse, A can be leaning to trust C as a good nurse. What does this means? This mean that C trusts A as for a given *competence* in a given *domain* [5]: B is considered by A an 'authority' in this domain, this is why A considers B's evaluation as highly reliable (B is a trustworthy source of evaluation).

So, since A delegates C the task τ we should have:

$$Inf_A\left(B\,C\,e(\tau)\right) \cdot DoT_{A\,B\,e(\tau)} > \sigma_1, \qquad (26)$$

where σ_1 is the A's reasonable trust threshold for delegating τ.

We have a pseudo-transitivity (we consider it as cognitively-mediated transitivity) when:

1) $Inf_A (B\,C\,e(\tau)) = 1$, in words, there is no influence on B by C performing τ (following A); and
2) $DoT_{A\,B\,e(\tau)} = 1$, in words, A has a perfect trust on B as evaluator of competence, opportunities, etc. in the task τ.

If the two above conditions are satisfied have $DoT_{A\,C\,\tau} = DoT_{B\,C\,\tau}$. More in general, given $Inf_A (B\,C\,e(\tau)) = 1$, each time $DoT_{B\,C\,\tau} > \sigma_2$ and $DoT_{A\,B\,e(\tau)} > \sigma_1$, we will also have $DoT_{A\,C\,\tau} > \sigma_1$ In Table 3 we show the various possibilities of A's trusting or not C combining the two DoTs.

Table 3.

	$DoT_{B\,C\,\tau} > \sigma_2$	$DoT_{B\,C\,\tau} < \sigma_2$
$DoT_{A\,B\,e(\tau)} > \sigma_1$	A trusts C $DoT_{A\,C\,\tau} > \sigma_1^*$	A does not trust C $(DoT_{A\,C\,\tau} < \sigma_1^*)$ or A trusts C $(DoT_{A\,C\,\tau} > \sigma_1^*)$
$DoT_{A\,B\,e(\tau)} < \sigma_1$	A trusts C or A does not trust C	A does not trust C $DoT_{A\,C\,\tau} < \sigma_1^*$

Conformism

This mechanism is not based on a special expertise or authority of the "models": they are not particularly expert, they must just have experience and trust: since they do I do; since they trust I trust.

The greater the number of people that trust the greater my feeling of safety and trust;

(less perceived risk)

the greater the perceived risk the greater the necessary number of 'models'.

A good example of this is the use of credit cards for electronic payment or similar use of electronic money. It is a rather unsafe procedure, but it is not perceived as such, and this is mainly due to the diffusion of the practice itself: it can be a simple conformity feeling and imitation; or it can be an explicit cognitive evaluation like: 'since everybody do it, it should by quite safe (apparently they do not incur in systematic damages)'. If everybody violates traffic rules, I feel encouraged to violate them too.

More formally, we have:

if Bel $(A\,(DoT_{BC\tau} > \sigma_2) \cap (DoT_{DC\tau} > \sigma_3) \cap$ .... $\cap (DoT_{ZC\tau} > \sigma_n))$ then $DoT_{AC\tau} > \sigma_1$. We can also say that in a conformism situation:

when $n \to \infty$ then $DoT_{AC\tau} \to 1$ and $\sigma_1 \to 0$.

A very crucial example of this trust-spreading mechanism is about circulating information.

The greater the number of (independent) sources of an information, a reputation, a belief, the more credible it is.

In fact belief acceptance is not an automatic process, it is subject to some test and check: one checks for its plausibility, coherence with previous knowledge, for the source reliability, etc. Thus, if several cognitive agents believe *belief1*, probably *belief1* has been checked by each of them against its own direct experience, previous knowledge, and source evaluation, thus it is reasonable that it is more credible. Even in science convergence is a criterion of validity. However, this is also a rather dangerous mechanism (of social bias and prejudice) since in fact sources in a social context are not independent and we cannot ascertain their independence, thus the number of sources can be just an illusion: there could be just an unique original source and a number of "reporters".

All these forms of trust spreading are converging in a give *target* of trust (being it an information source, a technology, a social practice, a company, a guy) (Fig.5(A)).

4.2   Since A Trusts B, (by Analogy) C Trusts D

This is a generalised form of trust contagion. It can be based on cognitive analogy or co-categorization:

- since A trusts B (and A is expert) and D is like B, C will trust D
  or
- since everybody trusts some B, and D is like/a B, C will trust D.

Where "like" either means "D is similar to B as for the relevant qualities/requirements", or means "D is of the same category of B"; and "some B" means "someone of the same category of B or similar to B".

Also in this case either a specific model and example counts, or a more generalized practice.

4.3   Calculated Influence

As we said, usually, this trust contagion is a macro-level phenomenon and the individual agent do not calculate it, it is not intended. However, sometimes it is a strategic decision. Suppose for example that A believes to be a "model" for C, and that he wants C trusts B (or D). In this case A can on purpose make C believe that he trusts B, in order to influence C and induce him to trust B (or D). The smart businessman can be aware of the fact that if they buy certain shares a lot of other people will follow him, and it can exploit precisely this imitative behavior of his followers for speculation at their expenses.

All the previous mechanisms (4.1 till 4.2) are responsible for that celebrated *"trust atmosphere"* that is claimed to be the basis of a growing market economy or of a working political regime. They are also very fundamental in computer mediated organizations, interactions (like electronic commerce), cooperation (CSCW), etc. and even in multi-agent systems with autonomous agents.

5   Concluding Remarks

Strategies and divices for trust building in MAS and vitual societies should take into account the fact that social trust is a very dynamic phenomenon both in the mind of the agents and in the society; not only because it evolves in time and has a history, that is A's trust in B depends on A's previous experience and learning with B itself or with other (similar) entities. We have in fact explained how *trust is influenced by trust* in several rather complex ways. In particular we have discussed three crucial aspects of such a dynamics and we have characterized some mechanism responsible for it and some preliminary formalization of it. We have modelled:

a) *How A's trusting B and relying on it in situation Ω can actually (objectively) influence B's trustworthiness in Ω.* Either trust is a self-fulfilling prophecy that modifies the probability of the predicted event; or it is a self-defeating strategy by negatively influencing the events. Both B's ability (for example through B's increased self-confidence) and B's willingness and disposition can be affected by A's trust or distrust and delegation. B can for example accept A's tacit exploitation and adopts A's goal, or negatively react to that. A can be aware of and take into account the effect of its own decision in the very moment of that decision. This makes the decision of social trusting more complicated and 'circular' than a trivial decision of relying or not on a chair.

b) *How trust creates a reciprocal trust, and distrust elicits distrust;* for example because B knows that A is now dependent on it and it can sanction A in case of violation; or because B believes that bad agents are suspicious and diffident (attributing to the other similar bad intentions) and it interprets A's trust as a sign of lack of malevolence. We also argued that the opposite is true: A's trust in B could induce lack of trust or distrust in B towards A, while A's diffidence can make B more trustful in A.

c) *How diffuse trust diffuses trust (trust atmosphere)*, that is how A's trusting B can influence C trusting B or D, and so on. Usually, this is a macro-level phenomenon and the individual agent do not calculate it. We focused on *pseudo-transitivity* arguing how indirected or mediated trust always depends on trust in the mediating agent: my accepting X's evaluation about Z or X's reporting of Z's information depends on my evaluation of Z's reliability as evaluator and reporter. In other terms this is not a logical or authomatic process but it is cognitively mediated.

We also discussed more basic form trust contagion simply due to diffusion of behaviours and to imitation because of a feeling of safety. Usually, this are macro-level phenomena and the individual agents do not calculate it.

References

[1] C. Jonker and J. Treur (1999), Formal analysis of models for the dynamics of trust based on experiences, *Autonomous Agents '99 Workshop on "Deception, Fraud and Trust in Agent Societies"*, Seattle, USA, May 1, pp.81-94.

[2] S. Barber and J. Kim, (2000), Belief Revision Process based on trust: agents evaluating reputation of information sources, *Autonomous Agents 2000 Workshop on "Deception, Fraud and Trust in Agent Societies"*, Barcelona, Spain, June 4, pp.15-26.

[3] A. Birk, (2000), Learning to trust, *Autonomous Agents 2000 Workshop on "Deception, Fraud and Trust in Agent Societies"*, Barcelona, Spain, June 4, pp.27-38.

[4] Castelfranchi C., Falcone R., (2000), Trust and Control: A Dialectic Link, *Applied Artificial Intelligence* journal, Special Issue on "Trust in Agents" Part1, Castelfranchi C., Falcone R., Firozabadi B., Tan Y. (Editors), Taylor and Francis 14 (8), pp. 799-823.

[5] Castelfranchi C., Falcone R., (2000). Trust is much more than subjective probability: Mental components and sources of trust, *32nd Hawaii International Conference on System Sciences* - Mini-Track on Software Agents, Maui, Hawaii, 5-8 January 2000.

[6] Castelfranchi C., Falcone R., (1998) Principles of trust for MAS: cognitive anatomy, social importance, and quantification, *Proceedings of the International Conference on Multi-Agent Systems (ICMAS'98)*, Paris, July, pp.72-79.

[7] Castelfranchi, C., Falcone, R., (1998) Towards a Theory of Delegation for Agent-based Systems, *Robotics and Autonomous Systems*, Special issue on Multi-Agent Rationality, Elsevier Editor, Vol 24, Nos 3-4, , pp.141-157.

[8] C. Castelfranchi, Social Power: a missed point in DAI, MA and HCI. In Decentralized AI. Y. Demazeau & J.P.Mueller (eds) (Elsevier, Amsterdam 1991) 49-62.

[9] Sichman, J, R. Conte, C. Castelfranchi, Y. Demazeau. A social reasoning mechanism based on dependence networks. *In Proceedings of the 11th ECAI*, 1994.

[10] Jennings. N.R. 1993. Commitments and conventions: The foundation of coordination in multi-agent systems. The Knowledge Engineering Review, 3, 223-50.

[11] Cohen, Ph. & Levesque, H., "Rational Interaction as the Basis for Communication". Technical Report, N89, CSLI, Stanford. 1987.

[12] Castelfranchi, C., Commitment: from intentions to groups and organizations. In *Proceedings of ICMAS'96*, S.Francisco, June 1996, AAAI-MIT Press.

[13] D. Gambetta, editor. Trust. Basil Blackwell, Oxford, 1990.

Belief Revision Process Based on Trust: Agents Evaluating Reputation of Information Sources

K. Suzanne Barber and Joonoo Kim

The Laboratory for Intelligent Processes and Systems
Electrical and Computer Engineering
The University of Texas at Austin
Austin, TX, 78722, USA
{barber, Joonoo}@lips.utexas.edu

Abstract. In this paper, we propose a multi-agent belief revision algorithm that utilizes knowledge about the reliability or trustworthiness (reputation) of information sources[1]. Incorporating reliability information into belief revision mechanisms is essential for agents in real world multi-agent systems. This research assumes the global truth is not available to individual agents and agents only maintain a local subjective perspective, which often is different from the perspective of others. This assumption is true for many domains where the global truth is not available (or infeasible to acquire and maintain) and the cost of collecting and maintaining a centralized global perspective is prohibitive. As an agent builds its local perspective, the variance on the quality of the incoming information depends on the originating information sources. Modeling the quality of incoming information is useful regardless of the level and type of security in a given system. This paper introduces the definition of the trust as the agent's confidence in the ability and intention of an information source to deliver correct information and reputation as the amount of trust an information source has created for itself through interactions with other agents. This economical (or monetary) perspective of reputation, viewing reputation as an asset, serves as social law that mandates staying trustworthy to other agents. Algorithms (direct and indirect) maintaining the model of the reputations of other information sources are also introduced.

Keywords: Multi-agent belief revision, trust management, reputation, Bayesian belief networks

1 Introduction

The term *trust* is increasingly used by many multi-agent researchers concerned with building real world multi-agent systems (MAS) that face partial, incomplete, uncertain, or even incorrect knowledge from diverse information sources.

---

[1] Information sources = {sensors, agents}. We also model the reliability of sensors. Although sensors do not have malicious intention, in real-world systems, they often give faulty information. The only difference between sensors and agents as information sources is that agents have intentions while sensors do not.

R. Falcone, M. Singh, and Y.-H. Tan (Eds.): Trust in Cyber-societies, LNAI 2246, pp. 73–82, 2001.

Since agents in these systems do not have complete knowledge or the ground truth about the problem domain, the agents often have different perspectives. To reduce the risk of deception and fraud in such systems, there have been efforts to infuse security into the infrastructure of the MAS [10; 17]. However, even in a secure MAS, it is frequently difficult, or even impossible, for an agent to know if an information source is unreliable due to maliciousness or just incompetence. Unreliable communication channels can act as culprits. Hence, in our research, we focus on modeling and maintaining credibility information about information sources which the agents utilize in generating more favorable behaviors, behaviors such as forming organizations or accepting information and revising its beliefs. We will assume some level of security is already incorporated at the infrastructure level.

Like many other terms in AI, many definitions for the term *trust* have been introduced. Reagle [15] categorizes trust into three groups:

1. "Trust as truth and belief": trust as confidence on quality/attribute of an entity or a statement.
2. "Trust as expectation": trust as the expectation of an assertion being true.
3. "Trust as commerce": trust as "confidence in the ability and intention of a buyer to pay at a future time for goods supplied without present payment."

We adopt but slightly modify the last category and define trust as confidence in the ability and intention of an information source to deliver correct information. *Information certainty* is defined as confidence on quality of a statement. This closely resembles the definition from the first category.

Reputation is closely related to trust. Our definition of reputation is the amount of trust an agent gives an information source based on previous interactions among them. If an information source consistently meets the expectations of other agents by delivering trustworthy information, other agents may increase the reputation of the information source. Likewise, not satisfying other agents' expectation due to either incompetence or maliciousness will decrease the information source's reputation among agents. The reputation of an information source S_1 is represented as $P(S_1^{reliable})$. It is a probability distribution where $P(S_1^{reliable}) + P(S_1^{unreliable}) = 1$ or simply $P(S_1^r) + P(S_1^u) = 1$. An information source is considered *reliable* if the agent receiving the information considers it trustworthy. Formally, an agent may internally model reliability of an information source as a threshold probability for error-free knowledge delivery for a specified time interval in a specified environment for a specific purpose.

Reputation could also be viewed as an asset, something that provides a monetary flow to its owner. Cronk [7] reports to "follow the example of animal behavior studies in seeing communication more as a means to manipulate others than as a means to inform them." In other words, most communication serves the purpose of social influence, defined as "change in one person's beliefs, attitudes, behavior, or emotions brought about by some other person or persons [14]." If we accept this premise then the reputation of an information source not only serves as means of belief revision under uncertainty, but also serves as social law that mandates staying trustworthy to other agents. Though an agent can send unreliable information to other agents or even lie, the agent risks the reputation it has been building among agents. Agents with consistently low reputations will eventually be isolated from the agent society since other agents will rarely accept justifications or arguments from agents with relatively low reputations. This isolation of unreliable information sources from the system is called "soft security" [13]. The algorithm proposed in this paper mandates soft security through reputations of information sources.

The remainder of this paper is organized as follows. The proposed belief revision process using reputation is explained in Section 2. Section 3 shows how an agent maintains reputations about other information sources. Finally, Section 4 summarizes the paper and addresses future research direction.

2   Belief/Model Revision through Reputation

Formalisms used for the representation of uncertain/incomplete information are often classified in two categories. In *logical formalisms*, such as AGM revision [1], KM update [11], Iterated Revision (IR) [5; 8], Transition-Based Update (TBU) [6], and Generalized Update (GU) [4], mathematical logic is extended to represent intuitive patterns of reasoning that involve incomplete/uncertain information. In *numerical formalisms*, such as approaches based on Dempster-Shafer belief functions, weights are associated with the pieces of knowledge, representing their degree of certainty. are good examples of numerical formalisms.

Dempster-Shafer (D-S) theory has been proven as a successful formalism for reasoning with uncertainty; unfortunately, D-S does not allow an information source to transmit it assessment of the quality of the knowledge it is transmitting. Our belief is that, by sending certainty information along with knowledge, the communication acts become more expressive and the receiver agent can support a more sophisticated belief revision process. For example, equipped with our algorithm, agent A can tell agent B that it is raining with certainty 0.9. D-S approaches like Dragoni's [9] can be thought of the special case, where the certainty is always 1.0. Since DS approaches implicitly assume that the information certainty is always 1.0, an agent cannot send knowledge with low certainty value. For instance, if Agent A believes it is raining with 0.4 certainty, it may not want to tell it other agents. Otherwise, it may jeopardize the reputation it has built in the agent society. Our algorithm allows an agent be able to explicitly convey/express its lack of confidence in this information.

In our algorithm, each agent maintains two types of belief bases; a background knowledge base (*KB*) and a working knowledge base (K). The KB contains knowledge an agent has accumulated. It is often inconsistent[2].

The K is the working memory an agent use to generate behaviors. It is a maximally consistent set of knowledge derived from the KB. An agent's reasoning and decision process are based on K.

---

[2] A model itself is consistent. The inconsistency in KB can result from the conflict in beliefs among models.

The proposed belief revision and model maintenance algorithm consist of the following steps:

1. Acquire knowledge q represented in terms of a propositional language L from information sources;
2. Build inference polytrees from the justifications accumulated for the given knowledge q;
3. Revise certainty factors of the sentences in KB (Combining evidence and updating belief node in the given polytrees[3]);
4. Generate the maximally consistent knowledge base K through certainty factor ordering;
5. (optional) Find counter-evidence and revise the reputations of the involved information sources;
6. (optional) Report conflicts back to the information sources.

Step 1

Suppose an information source S_1 sends knowledge q to agent X. The certainty factor that S_1 has on q is α , or S_1 thinks q is true with probability α. This communication act can be represented as $send$ (S_1, X, q, α) where S_1 is sender, X is receiver, q is the knowledge transferred, and α is the certainty the sender S_1 has on q, the knowledge transferred. At the end of this step, KB contains $\langle S_1, q, \alpha' \rangle$, which means the self-agent X believes q with certainty α' based on the evidence S_1 provided.

Step 2

Suppose n information sources, $S_1, S_2, \ldots, S_n$, have contributed to the current belief q so far and m information sources, $S_{n+1}, S_{n+2}, \ldots, S_{n+m}$, assert q. The parent nodes of the polytree consist of union of these information sources. Figure 1 shows resulting polytree.

$$\{S_1, S_2, \ldots, S_k\} = \{S_1, S_2, \ldots, S_n\} \cup \{S_{n+1}, S_{n+2}, \ldots, S_{n+m}\} \qquad (1)$$

The reputation for each information source and the certainty values they have on q come from KB.

Step 3

Once the polytree is built for q in Step 2, the certainty value X has on q, $P(q^{\text{true}})$ or simply $P(q^t)$, can be calculated by propagating probabilities in the tree. Every node in the tree has a binary value.

---

[3] A polytree is singly connected Directed Acyclic Graph (DAG).

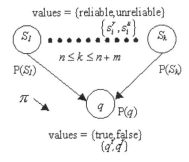

Fig. 1. The polytree with k information sources contributing to q.

Following [12][4] and assuming that the reliability of information sources is conditionally independent, probability propagation downward, π, from information sources to q is given by equation (2) with 2^π terms.

$$\pi(q^i) = \mathbf{C} \underset{l \in \{r, u\}}{} \prod_{m=1}^{k} P(q^i | s_m^l) \pi(s_m^l) \tag{2}$$

where $q^t \equiv (q^{true})$, $q^f \equiv (q^{false})$, $s^r \equiv (s^{reliable})$ and $s^u \equiv (s^{unreliable})$.
   This yields

$$\pi(q) = \pi(\pi(q^t), \pi(q^f)) \tag{3}$$

$P'(q^t)$, the revised certainty value for q in Agent X or the conditional probability of q^t based on the variables instantiated so far, is given by

$$P'(q^t) = \xi\pi(q^t) \tag{4}$$

where ξ is a normalizing constant so that $|P'(Q)| = 1$, or $P'(q^t) + P'(q^f) = 1$ (Bayesian conditioning). This is naive Bayesian approach in the sense that there may exist dependences among information sources. Although implicit dependencies among information sources may exist and thus violate independence assumption, various Bayes approaches have been shown to be resilient to these dependencies during operation. Empirical results for the approach presented in this paper have shown similar results with respect to possible dependencies among information sources.
   Agent X knows most of the parameters to calculate $P'(q^t)$ but $P'(q^t | s_i^u)$, the probability that the information it gives, q, is correct given the fact information source S_i is unreliable. Since KB does not contain knowledge to compute $P'(q^t | s_i^u)$, this algorithm employs the Random World assumption and $P'(q^t | s_i^u)$ can be estimated as $\mathbf{min}(\frac{1}{n_q}, P(s^r))$ where n_q is the number of possible values for q. For a binary random variable q, $n_q = 2$. If there is no conflicts in the steps forward, $\langle q, P'(q^r) \rangle$ enters into the working memory K.

_____
[4] The method proposed here is based on the treatment and the notations in [12].

Example 1

Suppose agent X just received $send(S_1, X, q, \alpha)$ from an information source S_1. X already has q in KB and justification is retrieved as $\langle S_2, Q, \beta \rangle$[5]. Figure 2 shows the polytree built in Step 2 from the given problem description.

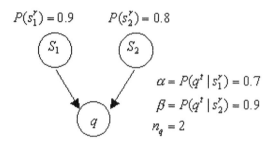

$P(s_1^r) = 0.9 \qquad P(s_2^r) = 0.8$

$S_1 \qquad S_2$

$\alpha = P(q^t \mid s_1^r) = 0.7$
$\beta = P(q^t \mid s_2^r) = 0.9$
$n_q = 2$

q

Fig. 2. Prior probabilities for example 1.

For the each information source S_i, we calculate message $\pi(S_i)$.

$$\begin{aligned}
\pi(S_1) &= (\pi(s_1^r), \pi(s_1^u)) = (P(s_1^r), P(s_1^u)) = (0.9,\ 0.1) \\
\pi(S_2) &= (\pi(s_2^r), \pi(s_2^u)) = (P(s_2^r), P(s_2^u)) = (0.8,\ 0.2)
\end{aligned} \tag{5}$$

Applying (5) for $\pi(q)$, we get

$$\begin{aligned}
\pi(q^t) &= P(q^t|s_1^r)P(q^t|s_2^r)P(s_1^r)P(s_2^r) + + P(q^t|s_1^r)P(q^t|s_2^u)P(s_1^r)P(s_2^u) + \\
&+ P(q^t|s_1^u)P(q^t|s_2^r)P(s_1^u)P(s_2^r) + + P(q^t|s_1^u)P(q^t|s_2^u)P(s_1^u)P(s_2^u) = \\
&= 0.7 \times 0.9 \times 0.9 \times 0.8 + 0.7 \times 0.5 \times 0.9 \times 0.2 \\
&+ 0.5 \times 0.9 \times 0.1 \times 0.8 + 0.5 \times 0.5 \times 0.1 \times 0.2 \\
&= 0.4635 + 0.063 + 0.036 + 0.005 = 0.5675
\end{aligned} \tag{6}$$

$$\begin{aligned}
\pi(q^f) &= P(q^f|s_1^r)P(q^f|s_2^r)P(s_1^r)P(s_2^r) + P(q^f|s_1^r)P(q^f|s_2^u)P(s_1^r)P(s_2^u) + \\
&+ P(q^f|s_1^u)P(q^f|s_2^r)P(s_1^u)P(s_2^r) + P(q^f|s_1^u)P(q^f|s_2^r)P(s_1^u)P(s_2^r) = \\
&= 0.3 \times 0.1 \times 0.9 \times 0.8 + 0.3 \times 0.5 \times 0.9 \times 0.2 + \\
&+ 0.5 \times 0.1 \times 0.1 \times 0.8 + 0.5 \times 0.5 \times 0.1 \times 0.2 = \\
&= 0.0216 + 0.027 + 0.004 + 0.005 = 0.0576
\end{aligned} \tag{7}$$

To find the normalizing factor ξ,

$$\xi = \frac{1}{0.5675 + 0.0576} = 1.600 \tag{8}$$

The revised certainty factor for q of X will be

$$P'(q^r) = \xi \pi(q^r) = 1.600 \times 0.5675 = 0.908 \tag{9}$$

---

[5] Information source S_2 gave knowledge q with certainty β previously

Step 4

Once all beliefs in the KB are assigned certainty values, set-covering [16] is applied to find K that is a maximally consistent set. If there is conflicting knowledge in KB, the knowledge with greater certainty enters into K. When there is conflicting knowledge with the same certainty value, we may (1) reject all of them or (2) pick one at random. To make K maximal, we select one at random.

Step 5

As a byproduct of the process in Step 3, Agent X can find a minimally inconsistent set of knowledge with respect to incoming knowledge q, if any. Among this set of conflicting beliefs the one with highest certainty value enters to K. Using this as evidence, an agent can update posterior probabilities of respective information sources S_i, $P(s_i^r)$ and $P(s_i^u)$.

Example 2: Suppose we have conflicting knowledge p and q and information sources S_1, S_2, and S_3. S_1 and S_2 support q, while S_3 supports p. Assume, from Step 3, we got $P(q^t) = 0.9$ and $P(p^t) = 0.5$. Since $P(q^t)$ is greater than $P(p^t)$, q enters to K. The evidences are $P'(q) = (1,0)$ and $P'(p) = (0,1)$. The upward probability propagation is calculated using upward message λ as

$$\begin{aligned} \lambda(q) &= P'(q) = (1,0) \\ \lambda(s_1^i) &= \sum_{j \in r,u} \pi(s_2^j)[\sum_{k \in t,f} P(q^k|s_1^i, s_2^j)\lambda(q^k)] \end{aligned} \tag{10}$$

$$\begin{aligned} \lambda(s_1^r) &= \pi(s_2^r)[P(q^t|s_1^r, s_2^r)\lambda(q^t) + P(q^f|s_1^r, s_2^r)\lambda(q^f)] + \\ &+ \pi(s_2^u)[P(q^t|s_1^r, s_2^u)\lambda(q^t) + P(q^f|s_1^r, s_2^u)\lambda(q^f)] = \\ &= \pi(s_2^r)[P(q^t|s_1^r)P(q^t|s_2^r)\lambda(q^t) + P(q^f|s_1^r)P(q^f|s_2^r)\lambda(q^f)] + \\ &+ \pi(s_2^u)[P(q^t|s_1^r)P(q^t|s_2^u)\lambda(q^t) + P(q^f|s_1^r)P(q^f|s_2^u)\lambda(q^f)] \end{aligned} \tag{11}$$

$$P(q^k|s_1^i, s_2^j) = P(q^k|s_1^i)P(q^k|s_2^j) \leftarrow s_1^i \; \underrightarrow{conditionaly \; indipendent} \; s_2^j \tag{12}$$

$$\begin{aligned} P(q^t|s_1^i) + P(q^f|s_1^i) &= 1 \\ P(q^t|s_2^i) + P(q^f|s_2^i) &= 1 \end{aligned} \tag{13}$$

Similarly we can compute $\lambda(s_1^u)$, $\lambda(s_2^r)$, $\lambda(s_2^u)$, $\lambda(s_3^r)$ and $\lambda(s_3^u)$. In computing $\lambda(s_k^i)$ $k = 1, 2, 3$, as in Step 3, Agent X knows all parameters but $P(q^f|s_k^u)$ and $P(q^t|s_k^u)$. X uses the same estimate $\mathbf{min}(\frac{1}{n_q}, P(s^r))$ as in step 3.

Finally, $P'(s_k^i)$, the revised conditional probability about the reputation of the information source S_k based on the evidence so far is given by

$$P'(s_k^i) = \xi\lambda(s_k^i)\pi(s_k^i) \text{ where } i \in \{r, u\}, k \in \{1, 2, 3\} \tag{14}$$

where ξ is normalizing constant assigned so that

$$|P'(S_k)| = 1 \text{ or } P'(ss_k^r) + P'(s_k^u) = 1. \tag{15}$$

Step 6

Although we are not enforcing global consistency of beliefs across all agents in a system, if a set[6] of agents is working on shared goals and they have an inconsistent set of knowledge supporting the goals, the overall performance of the group will be compromised. To prevent this, agent X may send its knowledge q supporting a goal g to agent Y when X finds Y has inconsistent and/or incorrect knowledge about g. The act can be represented as $send(X, Y, q, P(K_X q^t))$[7].

3   Reputation Management

In our system, there exist two paths through which an agent can acquire the reputation of other information sources:

1. *Indirect*: As in Step 6 of the belief revision process discussed in Section 3, when there are conflicts among acquired knowledge, an agent initiates the reputation revision process.
2. *Direct*: When communication is available, an agent can ask the other agents about the reputations of third party information sources. For example, agent X can ask agent Y about Y's belief on the reputation of agent Z. Agent X receives $\langle Y, R(Z), \alpha \rangle$ and performs belief revision process as in Section 3 to revise its belief on the reputation of Z.

The above distinction between direct and indirect reputation acquisition corresponds to Yahalom's distinction "between directly trusting some entity in some respect, and directly trusting an entity in some respect due to some other entity's trust [18]."

4   Summary and Future Work

We investigated a distributed belief revision algorithm for agents in real world multi-agent systems based on the reputation of related information sources. This research employs economical perspective to the trust and the reputation, which allows us to view the reputation as an asset, something that provides a monetary flow to its owner. This enforces individual agent "soft security" rather than hard security (infrastructure level security such as secure communication with public keys or authenticated name services).

Based on this framework, our belief revision algorithm provides efficient numerical treatment of inherent uncertainty due to the nature of practical application domain combining existing techniques such as Demster-Shafer theory of

---

[6] The group of agents working on a shared goal is called a team. Precisely speaking a team is concept based on goals. For example, in a complex agent system, agent X and Y can be teammates for goal A but adversaries for goal B. In Sensible Agent model, teams are called Autonomy Groups [2].

[7] $P(K_X q^t))$ is the certainty Agent X has on its belief q.

evidence combining and Bayesian belief networks. Our algorithm also is used to maintain underlying agent models. Direct and indirect reputation management algorithms for revising the reputation information of other information sources are also discussed.

Currently, we are collecting experimental data within the Sensible Agent Testbed [3] to evaluate the efficiency and effectiveness of our algorithms to maintain the models of other entities in the MAS. A possible limitation of our approach for computing the reputation of an information source could be the naïve Random World assumption to estimate $P(q^t|s^u)$, the prior conditional probability that knowledge q is correct given the originating information source S is unreliable. We will investigate performance gain by modeling it with past evidence since this additional maintenance will increase computational burden.

An expansion of the concept of reputation is also being considered. While we defined the reputation as a random variable, we can define it as an n-dimensional random vector that contains values such as (1) the current reputation the agent has (what we have now) and (2) the penalty that is the cost associated with abandoning the current trust (possibly by cheating). The penalty is related to the model of deception that is not covered in this paper. This extended model of reputation provides a means to mod el deception since when the gain from discarding the current reputation is greater than the penalty the agent might deceive others intentionally. Further research on the reputation model combining the concept of trust and deception will be followed.

Acknowledgements

This research was sponsored in part by The Texas Higher Education Coordinating Board, Grant #ATP 14-9717-0212, and DARPA TASK, Grant #26-0350-1612

References

[1]   Alchourrón, C., Gärdenfors, P., and Makinson, D.: On the Logic of Theory Change: Partial Meet Contraction and Revision Functions. Journal of Symbolic Logic, 50 (1985) 510-530.

[2]   Barber, K. S. and Martin, C. E.: Agent Autonomy: Specification, Measurement, and Dynamic Adjustment. In Proceedings of Autonomy Control Software Workshop at the Third International Conference on Autonomous Agents (Agents-99) (Seattle, WA, 1999) 8-15.

[3]   Barber, K. S., McKay, R. M., MacMahon, M. T., Martin, C. E., Lam, D. N., Goel, A., Han, D. C., and Kim, J.: Sensible Agents: An Implemented Multi-Agent System and Testbed. In Proceedings of Fifth International Conference on Autonomous Agents (Agents-2001) (Montreal, QC, Canada, 2001) 92-99.

[4]   Boutilier, C.: Generalized update: belief change in dynamic settings. In Proceedings of IJCAI '95 (1995) 1550-1556.

[5]   Boutilier, C.: Iterated Revision and Minimal Revision of Conditional Beliefs. Journal of Philosophical Logic, 25 (1996) 263-305.

[6]   Cordier, M.-O. and Lang, J.: Linking transitionbased update and base revision. In Proceedings of ECSQARU '95 (Fribourg, Switzerland, 1995) 133-141.

[7]   Cronk, L.: Communication as manipulation: Implications for biosociological research. In Proceedings of American Sociological Association Annual Meetings (Cincinnati, Ohio, 1991)

[8]   Darwiche, A. and Pearl, J.: On the Logic of Iterated Belief Revision. 89, 1-2 (1997) 1-29.

[9]   Dragoni, A. D. and Giorgini, P.: Belief Revision Through the Belief-function Formalism in a Multi-Agents Environment. In Proceedings of Intelligent Agents III, Agents Theories, Architectures, and Languages: ECAI'96 Workshop (Budapest, Hungary, 1996) Springer, 103-115.

[10] He, Q., Sycara, K. P., and Finin, T. W.: Personal Security Agents : KQML-Based PKI. In Proceedings of Second International Conference on Autonomous Agents (Minneapolis/St. Paul, MN, 1998) ACM Press, 377-384.

[11] Katsuno, H. and Mendelzon, A. O.: On the Difference between Updating a Knowledge Database and Revising It. In Proceedings of Second International Conference on Principles of Knowledge Representation and Reasoning (KR'91) (1991) Morgan Kaufmann, 387-394.

[12] Neapolitan, R. E. Probabilistic Reasoning in Expert Systems: Theory and Algorithms. Wiley, New York, 1990.

[13] Rasmusson, L. and Janson, S.: Simulated social control for secure Internet commerce. In Proceedings of New Security Paradigms '96 (1996) ACM Press.

[14] Raven, B. H.: A power/interaction model of interpersonal influence. Journal of Social Behavior and Personality, 7 (1992) 217-244.

[15] Reagle Jr., J. M. Trust in a cryptographic economy and digital security deposits: Protocols and policies. Master of Science Thesis, Department of Technology and Policy, MIT, 1996.

[16] Reggia, J. A., Nau, D. S., and Wang, P.: Diagnostic expert systems based on a set covering model. International Journal of Man-Machine Studies, 19, 3 (1983) 437-460.

[17] Wong, H. C. and Sycara, K.: Adding Security and Trust to Multi-Agent Systems. In Proceedings of Workshop on Deception, Fraud, and Trust in Agent Societies at The Third International Conference on Autonomous Agents (Agents-99) (1999) 149 - 161.

[18] Yahalom, R., Klein, B., and Beth, T.: Trust relationships in secure systems - A Distributed authentication perspective. In Proceedings of Proceedings of the 1993 IEEE Symposium on Research in Security and Privacy (1993) 153.

Adaptive Trust and Co-operation:
An Agent-Based Simulation Approach

Bart Nooteboom[1], Tomas Klos[2], and Renè Jorna[3]

[1] Erasmus University Rotterdam, the Netherlands
b.nooteboom@fbk.eur.nl
[2] TNO Institute of Applied Physics, the Netherlands
[3] Groningen University, the Netherlands

Abstract. Inter-firm relations have increasingly been analyzed by means of transaction cost economics (TCE). However, as has been widely acknowledged, TCE does not include dynamics of learning, adaptation or innovation, and it does not include trust. It assumes that efficient outcomes arise, while that may be in doubt, due to complexity and path-dependency of interactions between multiple agents that make, continue and break relations in unpredictable ways. We use the methodology of Agent-Based Computational Economics (ACE) to model how co-operation, trust and loyalty emerge and shift adaptively as relations evolve in a context of multiple, interacting agents. Agents adapt their trust on the basis of perceived loyalty. They adapt the weight they attach to trust relative to potential profit and they adapt their own loyalty, both as a function of realized profits. This allows us to explore when trust and loyalty increase and when they decrease, and what the effects are on (cumulative) profit.

Keywords: inter-firm relations, transaction costs, governance, trust, matching, complex adaptive systems, artificial adaptive agents, agent-based computational economics.
JEL Classification: C63, C78, D83, L22

1 Introduction

Inter-firm relations in general, and buyer-supplier relations in particular, have increasingly been analyzed by means of transaction cost economics (TCE). There are fundamental objections to that theory, some of which we will review below. However, in our view TCE also retains useful insights in relational risk of dependence (the 'hold-up' problem) and instruments for governing that risk, which are worth preserving in a wider theory of relations. Nooteboom (1992, 1996, 1999) has made attempts at such a wider theory, in a new synthesis. Here we proceed to apply that theory in a simulation model.

It has been widely acknowledged that TCE does not include dynamics of learning, adaptation or innovation. Focusing on governance, it has neglected competence. Williamson himself (1985, p. 144) admitted that 'the study of economic organization in a regime of rapid innovation poses much more difficult

R. Falcone, M. Singh, and Y.-H. Tan (Eds.): Trust in Cyber-societies, LNAI 2246, pp. 83–109, 2001.

issues than those addressed here'. A more fundamental problem is that as in mainstream economics more in general, it is assumed rather than investigated that efficient outcomes will arise. In inter-firm relations, it is assumed that optimal forms of organization or governance will arise, suited to characteristics of transactions such as the need for transaction-specific investments, frequency of transactions, and uncertainty concerning conditions that may affect future transactions (Williamson 1975, 1985). Two arguments are used for this: an argument of rationality and an argument of selection.

Williamson granted that rationality is bounded and transactions are subject to radical uncertainty, which precludes complete contingent contracting. But he proceeded to assume a higher level of rationality: people can rationally, calculatively deal with conditions of bounded rationality. They can assess the hazards involved and contract accordingly. However, if rationality is bounded, then rationality of dealing with bounded rationality is bounded as well. Under bounded rationality, how can one assume that agents correctly identify the hazards involved, as well as the effects of alternative modes of contracting? To rationally calculate economizing on bounded rationality, one would need to know the marginal (opportunity) costs and benefits of seeking further information and of further calculation, but for that one would need to decide upon the marginal costs and benefits of the efforts to find that out. This yields an infinite regress (Hodgson 1998, Pagano 1999). Here we accept bounded rationality more fully and deal with it on the basis of the methodology of adaptive agents. In a system of interacting agents without an all-knowing auctioneer, hazards depend on patterns of relations that are inherently unpredictable. In terms of the problem solving terminology of search, it is as if the fitness landscape (state space) in which one is searching is subject to earthquakes caused by oneself and others moving through it.

When confronted with arguments against rationality, economists usually concede that assumptions of full rationality are counterfactual, and then resort to the argument of economic selection. We can proceed as if agents make rational, optimal choices, because selection by forces of competition will ensure that only the optimal survives (Alchian 1950, Friedman 1953). Williamson was no exception in this respect. He held that in due course, market forces would select inefficient forms of organization out. However, that argument of selection has been shown to be dubious. For example, Winter (1964) showed that in evolution it is not the best *conceivable* but the best that happens to be *available* that survives. Due to effects of scale a large firm that is inefficient for its size may win out over efficient small firms. Furthermore, entry barriers may hamper the selection efficiency of markets. Koopmans (1957) concluded long ago that if the assumption of efficient outcomes is based on an argument of evolutionary process, its validity should be tested by explicitly modelling that process. Particularly in the study of inter-firm relations, we have to take into account the complexities and path-dependencies that may arise in the making and breaking of relations between multiple agents. That is what we aim to do in this article. As Coase (1998) recently admitted:

'[t]he analysis cannot be confined to what happens within a single firm. The costs of coordination within a firm and the level of transaction costs that it faces are affected by its ability to purchase inputs from other firms, and their ability to supply these inputs depends in part on their costs of co-ordination and the level of transaction costs that they face which are similarly affected by what these are in still other firms. What we are dealing with is a complex interrelated structure.'

Following up on Epstein and Axtell's (1996) suggestion, we let the distribution of economic activity across different organizational forms emerge from processes of interaction between these agents, as they adapt decisions to past experience. In other words, learning should be included. The system may or may not settle down, and if it does the resulting equilibrium may or may not be transaction cost economic. In any case, '[i]t is the process of becoming rather than the never-reached end points that we must study if we are to gain insight' (Holland 1992, p. 19). The methodology of artificial adaptive agents, in ACE, seems basically the right methodology to deal with this 'complex interrelated structure' of 'processes of interaction in which future decisions are adapted to past experiences'. We use that methodology to model interactions between firms, in the making and breaking of relations. For this, we need to specify, within the overall framework of ACE, a process of boundedly rational adaptation, based on a mutual evaluation of transaction partners that takes into account trust and profits. We begin with what we consider the simplest possible mechanism of adaptation, which is a simple re-inforcement model. In later development of the model, the learning process may be deepened to have more content of cognitive process. We model a system of buyer-supplier relations, because that best illustrates transaction cost issues.

We focus on the role of trust and loyalty, for two reasons. The first reason is that TCE does not incorporate trust, and this is an area where development of insight has priority. The second reason is that the central feature, in ACE, of adaptation in the light of experience seems particularly relevant to trust (Gulati 1995, Zucker 1986, Zand 1972).

Paragraph 2 deals with theory. Section 2.1 briefly characterizes the methodology of agent-based computational economics (ACE) that will be used. Section 2.2 discusses the issue of trust and opportunism. Section 2.3 discusses costs and profits of transactions. In Paragraph 3 we proceed to explain the design of our model. Section 3.1 indicates how buyers and suppliers are matched on the basis of their preferences, which include trust next to expected profit. Section 3.2 shows how we model costs and profits, Section 3.3 how we model trust and loyalty, and Section 3.4 how we model adaptation. Section 3.5 specifies the model, and section 3.6 discusses how the model is intialized or 'primed'. Paragraph 4 discusses a few preliminary experiments. The final Paragraph 5 summarizes conclusions and prospects for further research.

2   Theory

2.1   Agent-Based Computational Economics

Holland and Miller (1991) suggest to study economic systems as 'complex adaptive systems'. A complex adaptive system (CAS) 'is a complex system containing adaptive agents, networked so that the environment of each adaptive agent includes other agents in the system' (1991, p. 365). The methodology of ACE is a specialization of this to economics. This approach is used more and more often to study problems in economics, such as in the repeated prisoner's dilemma (Klos 1999a, Miller 1996, Stanley et al. 1994), in final-goods markets (Albin and Foley 1992, Vriend 1995), stock markets (Arthur et al. 1997), industrial markets (Péli and Nooteboom 1997), and labour markets (Tesfatsion 2000).

The essence of this approach is that economic phenomena are studied as they emerge from actual (simulated) interactions between individual, boundedly rational, adaptive agents. They are not deduced from abstract models employing representative agents, auctioneers or anonymous, random matching, etc. Rather, whether an interaction takes place between any two given agents is left for them to decide. What the agents subsequently do in that interaction is their own-possibly sub-optimal-decision, that they make on the basis of their locally available, incomplete information and as a result of their own (cognitively limited) processing of that information. Appropriate forms of reasoning are induction and abduction, rather than deduction as used in optimization models that are solved for 'never reached end points'.

2.2   Opportunism and Trust

ACE can be used to gain insight in the assumptions of economic theories, especially arguments that are used for the assumption of opportunism and the neglect of trust. It is instructive to analyze in some detail how TCE deals with opportunism and trust. A first argument is that since at the beginning of a relation one has no information about a partner's trustworthiness, one must take the possibility of opportunism into account and construct safeguards for it. Nooteboom (1999) argued that this involves an inconsistency concerning the time dimension in TCE. Time is essential for the central notion of transaction-specific investments. Such investments have to be made 'up front', while they can be recouped only in a succession of transactions in the future. When the relation breaks prematurely, there is a loss of specific assets. In the course of time there may be cumulative gains of efficiency due to learning by doing. When the relation breaks this advantage is lost. Both yield switching costs, and these create dependence, and thereby vulnerability to 'hold-up'. However, if allowance thus has to be made for the passage of time, in a relation, then one should also allow for the acquisition of knowledge concerning he trustworthiness of a transaction partner.

A second argument in TCE is that while it might be possible to gather information about a partner's trustworthiness, it is costly and imperfect. It is more efficient to accept the possibility of opportunism and construct safeguards against it. Here also an important point is assumed rather than analyzed. Safeguarding against opportunism may be more costly or difficult than gathering information about trustworthiness. The assumption of opportunism signals distrust, which may engender distrust on the other side of the relation, which can yield a vicious circle of regulation and constraint which reduces the scope for collaboration (Zand 1972).

A third argument against incorporating trust in TCE, offered by Williamson (1993), is that if trust does not go beyond calculative self-interest it adds nothing, and if it does it yields blind trust, which is unwise and will not survive. So trust should be left out. Against this, Nooteboom (1999) argued that trust can be real, in the sense that it goes beyond calculative self-interest, without being blind, because it can be bounded rather than unconditional. One can assume that people will be loyal only within constraints, i.e. below some limit that represents their resistance to temptations of opportunism.

In our simulation model we take trust and its limits into account in the following manner:

1. Trust is adaptive: it is based on experience with a (potential) partner's commitment to a relation, as exhibited by his past lack of defection to more attractive alternatives.
2. Trust goes beyond potential profit: it is included, next to profit, in the determination of a (potential) partner's attractiveness. The weight attached to trust relative to expected profit is also adaptive: it depends on realized profits.
3. Trustworthiness, in the sense of commitment to a relation, is limited: there is a threshold of resistance to temptation below which the agent will not defect to a more attractive alternative.
4. Trustworthiness also is adaptive: the threshold of defection is adjusted as a function of realized profits.

This implementation allows us to experiment with the development of trust and trustworthiness, the role this plays in making and breaking of relations, the resulting durability or volatility of relations, and the outcome regarding efficiency, in terms of realized profits. These experiments are conducted under different conditions concerning costs and profits that are relevant for transaction cost analysis. The specification of costs and profits is discussed in the next section.

2.3   Costs and Profits

A central concept in transaction cost analysis is the notion of 'transaction specific investments'. These yield profits as well as costs. The profit lies in differentiated products, which yield a higher profit than standardized products. With standardized products one can compete only on price, and under free and costless entry to the market this will erode price down to marginal cost, as proposed in standard microeconomics. In contrast, differentiated products allow for a profit margin.

To the extent that assets are not specific, they may be used for alternative partners and generate economy of scale. When production technology and/or marketing are inflexible, so that a specific product entails a specific investment, differentiated products entail specific investments (Nooteboom 1993). This entails loss of economy of scale. As analyzed in TCE, a specific investment has to be made 'up front' and is recouped only after repeated transactions, while it has less or no value in other applications, in transactions with another partner. Thus it entails switching costs in the form of loss of investment (and the need to make new investments for a new partner). A second type of switching cost arises from learning by interacting experience (or 'learning by doing', cf. Yelle 1979): as partners make specific investments in each other, they learn cumulatively. When the relation breaks, this advantage is surrendered. Both types of switching costs make one dependent on the commitment of the partner to continue the relation. That yields the 'hold-up' problem. In the present version of our model only this second form of switching costs (learning by doing) is implemented.

These features of profit from differentiated products, economy of scale, economy of experience and switching costs are incorporated in the simulation model. This enables us to explore the advantages and disadvantages of exclusive, durable relations versus non-exclusive, volatile ones. The next section discusses the design of the model.

3   The Model

3.1   Preferences and Matching

Rather than rely on standard, anonymous random matching devices, matching on the basis of the choice of partners is explicitly incorporated in the model. Agents are assumed to have differential preferences for different potential trading partners (cf. Weisbuch et al. forthcoming). On the basis of preferences, buyers are either assigned to suppliers or to themselves. When a buyer is assigned to himself this means that he 'makes rather than buys'. In other words: we endogenize the 'make or buy decision'. This process is generated by a so-called matching algorithm[1]. The current section describes the algorithm.

A matching algorithm produces a set of matches (called 'a matching') on the basis of individual agents' preference rankings over other agents. We express preferences in 'scores'. Each agent assigns a score to all agents it can possibly be matched to (where a buyer can be matched to himself, to make rather than buy). The score is a function of (1) the profit the agent can potentially make as a result of the transaction and (2) his trust in the potential partner. In line with the literature on trust (Gambetta 1988), trust is interpreted as a subjective probability that no harm will occur and that expectations will be fulfilled. We

---

[1] See (Roth and Sotomayor 1990) for an excellent introduction to and overview of matching theory.

use a multiplicative specification of potential profit and trust, to express that they interact: the score must be zero if either trust or potential profit is zero. The product of potential profitability and trust interpreted as a probability of realization would constitute expected profit. In order to allow agents to attach varying weights to profitability versus trust, instead of simple multiplication of the two we employ a Cobb-Douglas functional form:

$$\text{score}_{ij} = \text{profitability}_{ij}^{\alpha_i} \cdot \text{trust}_{ij}^{1-\alpha_i} \tag{1}$$

where: score_{ij} is the score i assigns to j, $\text{profitability}_{ij}$ is the profit i can potentially make 'through' j, trust_{ij} is i's trust in j and $\alpha_i \in [0,1]$ is the importance i attaches to profitability relative to trust, i.e. the 'profit-elasticity' of the scores that i assigns; i may adapt the value of α_i from each timestep to the next. In later sections we describe how profitability and trust are determined, and how adaptation takes place. Besides a preference ranking, each agent maintains a 'minimum tolerance level' that determines which partners are acceptable. Each agent also has a maximum number of matches it can be involved in at each time period (a quota).

The algorithm used for matching is Tesfatsion's (1997) deferred choice and refusal (DCR) algorithm, which extends Gale and Shapley's (1962) deferred acceptance algorithm. However, the DCR algorithm is used with some qualifications. First of all, and most importantly, unlike the DCR algorithm we do allow buyers to be matched to themselves, in which case they are their own supplier. Secondly, only disjoint sets of buyers and suppliers are allowed, so that there are no agents that can be buyer as well as supplier. So, although buyers may be their own supplier, they can not supply to other buyers. Finally, we allow different agents to have different quotas-i.e. different maximum numbers of matches allowed at any moment in time-because different buyers and suppliers are likely to want different numbers of partners. Each buyer includes itself as one of the alternatives in its preference ranking, and suppliers not ranking higher than the buyer are unacceptable. In other words: a buyer prefers to remain single (and 'make') rather than 'buy' from an unacceptable supplier.

The matching algorithm proceeds as follows. Buyers may have one or more suppliers and suppliers may have one or more buyers; each buyer i has an offer quota o_i and each supplier j has an acceptance quota a_j. Before the matching, all buyers and suppliers establish a strict preference ranking over all their alternatives. The algorithm then proceeds in a finite number of steps.

1. In the first step, each buyer sends a maximum of o_i requests to its most preferred, acceptable suppliers. The algorithm structurally favours the agents that send the requests; it is plausible that buyers do this. Because the buyers typically have different preference rankings, the various suppliers will receive different numbers of requests.

2. The suppliers first reject all requests received from unacceptable buyers[2]. Then, each supplier 'provisionally accepts' a maximum of a_j requests (acceptance quota) from its most preferred acceptable buyers and rejects the rest (if any).

3. Each buyer that was rejected in any step fills its quota oi in the next step by sending requests to (o_i minus the number of outstanding, provisionally accepted, requests) next-most-preferred, acceptable suppliers that it has not yet sent a request to.

4. Each supplier again rejects requests received from unacceptable buyers and provisionally accepts the requests from a maximum of a_j most preferred, acceptable buyers from among newly received and previously provisionally accepted requests and rejects the rest. As long as one or more buyers have been rejected, the algorithm goes back to step 3. The algorithm stops if no buyer sends a request that is rejected. All provisionally accepted requests are then definitely accepted.

3.2   Modelling Potential Profit

A buyer's potential to generate profits is a function of the buyer's position on the final market – where he is a seller – defined as the degree of product differentiation. A supplier's potential to generate profits is determined by the supplier's efficiency in producing for the buyer. The model allows for varying degrees of product differentiation. As indicated before, a more differentiated product yields a higher profit margin. This is expressed in a buyer-specific variable di (in the interval [0, 1]) that determines the profit the buyer will make when selling his products. We experiment with different values for d_i to see how they affect the choices that agents make.

Production, whether conducted by a supplier or by the buyer himself, requires investments in assets. One unit of asset is normally required to produce one product, but increasing efficiency may decrease this amount. Assuming that the production technology is more or less rigid (differentiated products require specialized assets), we assume a connection between the differentiation of a buyer's product and the specificity of the assets required to produce it. To the extent that assets are specific, they entail switching costs. On the other hand, if products are not differentiated, investments to produce the product for one buyer can easily be switched to producing the product for other buyers. The simplest way to model this relation is to assume that asset specificity is equal to product differentiation, i.e. the proportion of the asset that is specific to a buyer is equal to the extent to which that buyer's product is differentiated. If a buyer produces for himself, it makes no sense to distinguish between buyer-specific and non-specific assets. A buyer's minimum acceptance level of suppliers is the score that the buyer would attach to himself. Since it is plausible that he completely

---

[2] For the moment, we assume that all buyers are acceptable to the suppliers; suppliers do not, like the buyers, have any alternative, so they will rather supply to any buyer than remain single.

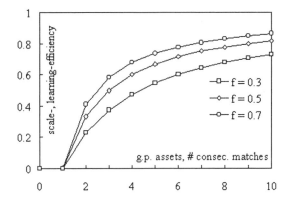

Fig. 1. Efficiency of scale and of learning-by-doing.

trusts himself, trust is set at its maximum of 1, and the role of trust in the score is disregarded: $\alpha = 1$.

If a supplier produces for one or more buyers, then his investments are split into two categories: buyer-specific and non-specific-i.e. general-purpose-assets. As explained above, the percentage of investment that is specific to that buyer is the same as the extent to which that buyer's product is differentiated. The supplier adds the remaining, general-purpose investment for each buyer over all the buyers he is matched to. The corresponding volume of production is subject to economy of scale. The utilization of specific assets is subject to experience effects: uninterrupted use for the buyer involved yields learning by doing. The scale and learning effects are modelled according to a curve illustrated in Figure 1, which plots the function:

$$y = max\left(0, 1 - \frac{1}{f_x + 1 - f}\right) \tag{2}$$

where y is scale-efficiency viz. learning efficiency; x is general-purpose assets viz. number of consecutive matches, and f is a scaling parameter.

This function is just a convenient way of expressing decreasing returns for both scale and experience effects. For the scale effect, it shows positive values along the vertical axis only for more than 1 general-purpose asset. This expresses that a supplier can be more scale-efficient than a buyer producing for himself only if the scale at which he produces is larger than the maximum scale at which a buyer might produce for himself. For the learning effect, a supplier's buyer-specific efficiency is 0 in their first transaction, and only starts to increase if the number of transactions exceeds 1.

The way profits are made, then, is that suppliers may reduce costs by generating efficiencies for buyers, while buyers may increase returns by selling more differentiated products. It is assumed that the profit that is made resulting from both partners' contributions is shared equally between the buyer and the supplier involved.

In the current model, the costs of investments are not made explicit, and are assumed to be included in the profit margin. As a result, the model does not include switching costs in the form of the loss of specific investments when a relation breaks. It only includes switching costs in the form of loss of the cost reductions concerning specific investments that result from learning by doing. That yields an equal opportunity cost for buyers and suppliers. This constitutes an important limitation of the model. The reason for it was that to make investment costs explicit one would need to specify the height of investment relative to net profit, the period in which the investment is amortized, and whether this is done linearly or exponentially. That yields a considerable complication. Nevertheless, it should be incorporated in future versions of the model.

3.3   Modelling Trust

As indicated above, trust is associated with a subjective probability that potential gain will indeed be realized. This interpretation of the notion of trust as a subjective probability, relating it to the notion of risk concerning the realization of intentions, is standard in the literature on trust (Gambetta 1988).

We focus on the risk that a partner will defect and thereby cause switching costs. Therefore, as noted before, in the score that a buyer attaches to himself, for in-house production as an alternative to engaging a supplier, we assume maximum trust, and the weight attached to trust $(1 - \alpha)$ is zero. In other words trust is 100%, but it is not relevant. Evaluating himself, the producer looks only at potential profit. In other words, we assume that firms will not be opportunistic to themselves. We make the subjective probability adaptive as a function of the defection behaviour of partners that one has experienced. This modelling of adaptive trust is cognitively quite meagre. In future developments of the model the underlying cognitive process of expectations should be developed.

Following Gulati (1995), we assume that trust increases with the duration of a relation. As a relation lasts longer, one starts to take the partner's behaviour for granted, and to assume the same behaviour (i.e. commitment, rather than breaking the relation) for the future (cf. the notion of 'habituation' in Nooteboom et al. 1997). In the trust literature, it has been argued that when a relationship proceeds in time, a process of mutual identification develops, in which partners are able to understand each other better and to engage in mutual forbearance and 'give and take' (Lewicki and Bunker 1996). As in the earlier discussion of effects of scale and experience, we assume that this is subject to decreasing returns to scale, and we use the same basic functional form employed before. Here, we add a base level of trust at the beginning of a relationship. This reflects the notion that prior to relations there is a basic, ex ante trust, as an institutional feature of a society (Nooteboom 1999, 2002). It may be interpreted as the expected percentage of non-opportunistic people, or as some standard of elementary decency that is assumed to prevail. On top of that basic level of trust one can develop partner-specific trust on the basis of experience in dealings with him, and mutual indentification. The model is specified as follows:

$$\text{trust} = b(1 - b)\left(1 - \frac{1}{f_x + 1 - f}\right) \tag{3}$$

In this formula, x refers to the uninterrupted duration of a relationship. The formula is similar to formula (2). It includes the same mechanism of increase subject to decreasing returns to scale, which is here added to the base level of trust (b).

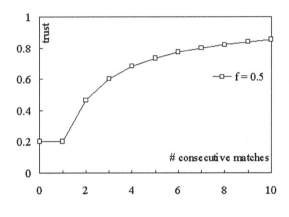

Fig. 2. Trust.

Figure 2 shows the relation between the past duration of a relation and agents' trust in each other for an agent-specific value of 0.5 for the f-parameter.

For each agent, trust starts out at a certain starting value that is set exogenously, and is adapted to experience with the loyalty of partners. If an agent i, involved in a relation with an agent j, 'breaks' their relation, then j's trust in i decreases. j's trust drops by a percentage of the distance between the current level and the base-line level of trust; it stays there until the next time j and i are matched, after which is starts to increase again for as long as the relation lasts without interruption. This discontinuous drop of trust when defection occurs reflects the saying that 'trust comes on foot and leaves on horseback'.

The other side of the coin is, of course, one's own trustworthiness. This is modelled as a threshold τ for defection. One defects only if the advantage over one's current partner exceeds that threshold. This entails an important theoretical move. It reflects the recognition that trustworthiness has its limits, and that trust should recognise this and not become blind (Pettit 1995, Nooteboom 2002). There is a limit to the resistance to temptations and pressures of opportunism. For firms, this depends on competitive pressure. For sheer survival, a firm may need to utilize every opportunity for profit that it can get. The threshold is also adaptive, as a function of realized profit. Technically, in the matching algorithm the threshold is added to the score for the current partner.

3.4   Adaptation

An agent in a CAS is adaptive if 'the actions of the agent in its environment can be assigned a value (performance, utility, payoff, fitness, or the like); and the agent behaves in such a way as to improve this value over time' (Holland and Miller 1991, p. 365). The adaptive character of the artificial agents in the present model refers to the possibility for the agents to change the values they use for α and τ from one timestep to the next. As discussed, a is the profit elasticity of the preference score ($1 - \alpha$ is the elasticity with respect to trust), and τ is the threshold for one's own defection. To each value of a and τ , each agent assigns a strength, which is its selection probability[3].

This expresses the agent's confidence in the success of using that particular value; the various strengths always add up to a constant C. At the beginning of each timestep each agent chooses a value for α and τ. In our model, the choice between the different possible values is probabilistic – a simple roulette wheel selection – with each value's selection probability equal to its relative strength, i.e. its strength divided by C. The strength of the value that was chosen at the start of a particular timestep, is updated at the end of that timestep, on the basis of the agent's performance during that timestep, in terms of realized profit[4]. As an output of the simulation, each agent is weighted average value is calculated. This indicates where i's emphasis lies: because the value with the highest strength pulls the weighted average in its direction, the emphasis lies on low values if the weighted average is low and vice versa.

3.5   The Simulation Model

After describing the logic of different parts of the model, concerning the matching mechanism, the modelling of costs, profits and trust, and the modelling of adaptation, we now show how they fit together in the simulation model. The simulation proceeds in a sequence of time steps, called a 'run'. Each simulation experiment may be replicated several times (multiple runs), to reduce the influence of draws from random distributions on the results. At the beginning of a simulation starting values are set for certain model parameters. The user is prompted to supply the number of buyers and suppliers, as well as the number of runs, and the number of timesteps in each run[5].

---

[3] This form of adaptation is called reinforcement learning. A classifier system is used to implement it. See (Arthur 1991, Arthur 1993, Kirman and Vriend 2000, Lane 1993) for discussions and other applications in economic models; good general introductions to classifier systems are (Booker et al. 1989), (Goldberg 1989) and (Holland et al. 1986).

[4] This works as follows: the agent adds the profit obtained during timestep t to the strength of the value that was used for α or τ. Then, all strengths are renormalized to sum to C again (see (Arthur 1993) for a discussion of this learning mechanism).

[5] The original simulation was developed in the general-purpose, object-oriented programming language SIMULA (Birtwistle et al. 1973). A new version with the same functionality and a graphical user interface was implemented in Java by Martin

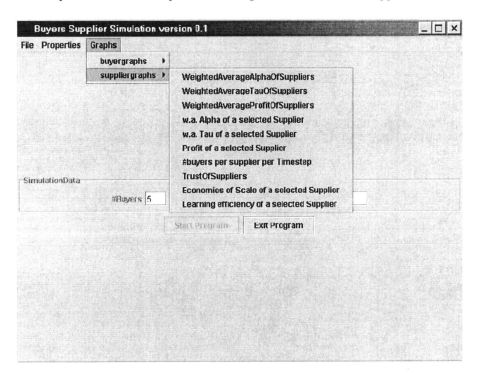

Fig. 3. The simulation program's graphical interface.

Figure 3 shows the interface of the program, where a choice can be made about which outputgraphs to show. Also, parameters of the simulation can be set.

The program's random number generator is seeded and finally, the agents are instantiated and given a number for identification. At the start of each run, each of the agents is initialized. For example, the agents' profits (from the previous run) are re-set to zero and the agents' trust in other agents is re-set to the base level. The simulation is described in more detail in Klos (1999b). The flow diagram in Figure 4 gives an overview.

Each timestep starts with a calculation of scores as a basis for matching. Each agent calculates scores and ranks potential partners accordingly. Random draws are used to settle the ranking of alternatives with equal scores. To calculate scores each agent chooses a value for a: the elasticity of the score with respect to potential profit. Potential profitability depends on profit margin, related to the degree of product differentiation, economies of scale and economies of experience. As discussed before, suppliers enjoy scale-economies in the total of general-purpose assets used in the production for multiple buyers. Furthermore, as a supply relation lasts, the supplier accumulates experience and efficiency at using specific

Helmhout. The object-oriented paradigm is very well suited for agent-based modelling (see McFadzean and Tesfatsion 1999, Epstein and Axtell 1996).

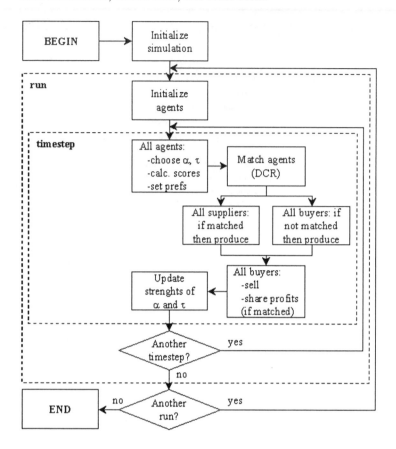

Fig. 4. Flow diagram of the simulation.

assets in the production for a particular buyer. Suppliers' scale-effciency is inferred from the outcome of the previous timestep. Only after the matching it becomes clear to how many and which buyers each supplier is actually matched, and what the real extent of his scale-efficiency is. Expectations of the supplier's position on each buyer-specific experience curve, on the other hand, will already be accurate before the matching-assuming, of course, that the relation makes it through the matching. After the calculation of scores the matching algorithm is applied. The outcome is determined by the agents' preference rankings over acceptable alternatives.

After the matching, suppliers that are matched to a buyer produce for their buyer(s), while buyers that are 'self-matched' (not matched to a supplier) produce for themselves. Assets that suppliers invest in for the production for a certain buyer are specific to that buyer to the extent that the buyer's product is differentiated. The remainder of the assets is 'general purpose'. After production, all buyers sell their products on the final-goods market. Net revenue is determined by the profit rate, which depends on the degree of product differentiation, and costs, depending on efficiency of scale and of experience. Net profits are equally divided between buyer and supplier. The events in this latter part of each timestep-i.e., after the matching-may lead the agents to adapt their preference rankings, used as input for the matching algorithm in the next timestep. In particular, they adapt their values of α and τ, as discussed in the previous section. A relation is broken if, during the matching, a buyer does not send any more requests to the supplier or he does, but the supplier rejects them. Across timesteps realized profits are accumulated for all buyers and suppliers, and all the relevant parameters are tracked.

3.6   Priming the Model

The simulation model did not come about instantly; it was developed over the course of a period of time. During the development of the model, results were constantly generated, and conclusions from those results were fed back into the development of the model. From the perspective of the current version of the model, this has resulted in a number of decisions concerning the construction of the model and the setting of parameters. Apart from giving interesting insights into the model and what it represents in its own right, this 'sensitivity analysis' has led to the parameter setting for an initial set of experiments as listed in Table 1.

Each simulation run involves 12 buyers and 12 suppliers and lasts for 500 timesteps. At several locations in the model, numbers are drawn from pseudo-random number generators. For example, such draws are necessary, in the matching algorithm, to settle the ranking of alternatives to which equal scores are assigned[6]. Another example is that, in selecting values for α and τ, each agent needs to draw two pseudo-random numbers. It would also have been possible to simply pick the value with the largest strength, but that would leave the agents too little room for exploration, because the first value chosen is then chosen forever after.

Pseudo-random numbers have an effect on the results. In order to diminish this influence, each run is repeated a number of times (25), and results are (sometimes) presented as averages across those 25 runs (and across all agents). In other cases, the focus is on the individual run or on the characteristics of the individual agents involved. Selection strengths are increased with realized profit, among others, which depends on product differentiation. The parameters were set so that agents did get the opportunity to explore, and were not locked in to the region in the problem space they were plugged into at the start of the simulation.

Also, only values between 0 and 0.5 were allowed for τ (the defection threshold), because higher values had too high an influence on the results. If there are

---

[6] Using the numbers that identify agents to solve this problem would lead to agents identified by lower numbers to gain a systematic (dis-)advantage over agents identified by high numbers, so the assignment is randomized.

Table 1. Parameters and variables in the simulation.

	param./var.	Value range	Value used
general	number of buyers, I	$\{1, 2, ...\}$	12
	number of suppliers, J	$\{1, 2, ...\}$	12
	number of runs, R	$\{1, 2, ...\}$	25
	number of timesteps, T	$\{1, 2, ...\}$	500
	number of values for $\alpha \in [0, 1]$	$\{2, 3, ...\}$	5
	number of values for $\tau \in [0, 0.5]$	$\{2, 3, ...\}$	5
	renormalization constant, C_α	$\langle 0, ... \rangle$	$3 \cdot d_i$
	renormalization constant, C_τ	$\langle 0, ... \rangle$	$3 \cdot d_i$
	baseTrust, b	$\langle 0, 1]$	0.3
	initTrust(subject)	$\langle 0, 1]$	0.75
	trustFactor	$[0, 1]$	0.5
per buyer i	differentiation, d_i	$[0, 1]$	$\{0.25, 0.35, ..., 0.75\}$
	offer quota, o_i	$\{1, 2, ..., J\}$	1
per supplier j	acceptance quota, a_j	$\{1, 2, ..., I\}$	3
	scaleFactor	$[0, 1]$	0.5
	learnFactor	$[0, 1]$	0.5

5 possible values between 0 and 1, then choosing either one of the three highest values results in a score-advantage of the current partner of at least 0.5. This is an advantage that alternative partners can almost never surpass, which means that agents very often stick to their first partner, into which situation the simulation then locks in. Using only values for τ between 0 and 0.5 gives the agents room to experiment with different values without getting locked into their initial situation.

Furthermore, each agent's initial trust in another agent was set at 0.75. It needs to be this high, because otherwise suppliers can never be more attractive than a buyer considers himself. Initially, suppliers enjoy no economies of scale or experience, so buyers have to be attracted to them by trusting them highly and by weighting profitability relatively low. If initial trust is not set high enough, buyers never prefer suppliers and nothing ever happens in the simulation. It could be argued that the buyers should foresee suppliers' scale economies and have a higher preference for them on that basis. The observation above has wider implications, however. The issue is that the advantages of the market in terms of scale-economies that TCE assumes do not come about instantly and only under certain circumstances. Time has to be allowed for these advantages to build up and this observation also forces one to allow for the fact that sometimes they do not even build up at all. More generally, the study of economic systems at the level of individual agents and the relations among them focuses attention on the way these kinds of micro-motives influence macro-behaviour (Schelling 1978).

Finally, each buyer's offer quota was fixed at 1, and each supplier's acceptance quota was set to 3. This means that each buyer has a supplier or he does not, in which case he supplies to himself, and that each supplier can supply to a maximum of 3 buyers. In previous experiments with each supplier j's acceptance quota set to the total number of buyers, the system quickly settled in a state where all buyers buy from the same supplier. For this reason, suppliers were only allowed a maximum of three buyers. This limits the extent of the scale economies that suppliers can reach.

4 Experiments

4.1 Make or Buy

In earlier experiments with the model (Klos and Nooteboom 2001) we varied the degree to which investment in any given buyer-supplier relation is specific, and we tested hypotheses concerning the effects on the 'make or buy' decision. Our hypothesis, taken directly from TCE, was that more product differentiation will favour 'make' over 'buy'. The model reproduced that outcome. This outcome thus confirms one claim of TCE.

One of our core objections to TCE was that it neglects problems that may obstruct the achievement of optimal outcomes. We investigated the effect on outcomes in terms of cumulative profit, to see to what extent optimal profits are indeed realized, and how this varies across runs of the model. Lack of optimal outcomes and a variety of outcomes would form an illustration of unpredictability and path-dependence in the formation of relations among multiple agents. Indeed, the simulations illustrate that optimal results are seldom achieved. Interestingly, optimal results are approached more closely when products are more differentiated, and, correspondingly, assets are more specific. The reason for this is that then buyers go for more 'make' rather than 'buy'. Then they are less subject to the uncertainties involved in which suppliers they will wind up with, and at what profit. This adds an insight that TCE was unable to give. It simply assumed that efficiency would be reached, in spite of its assumption of bounded rationality. Our model demonstrates how the uncertainty resulting from the complexity of interacting agents can block access to efficient outcomes. Here we can conclude that next to the classic advantage of 'make' over 'buy', for the control of hold-up risk, there now appears the additional advantage. With more 'make' and only incidental 'buy', there is less complexity of interaction, yielding fewer unpredictabilities and path-dependencies that make it difficult to achieve optimal results.

4.2 The Development of Trust

In the present paper we focus on our second core objection to TCE: the disregard of trust. Here we conduct experiments concerning the development of trust and trustworthiness. The expectation from TCE presumably is that trust does not pay, and then we would expect, in the adaptive process, a more or less uniform decline of both loyalty and the weight attached to trust. Our counterhypothesis is that to the extent that the advantages of durable 'buy' relations are higher, trust and trustworthiness, in terms of commitment to an ongoing relationship in spite of more attractive alternatives, matter more. We would expect that adaptive agents then evolve to relatively high levels of trustworthiness, less frequent switching, higher perceived commitment and hence trust, and a high weight attached to trust in the evaluation of partners. Below we present the evidence from the simulation.

A lot of data is generated in each simulation run. After the simulation, these data are used in a different application for generating the figures below; in addition, the simulation program provides a way of visualizing micro-level data, as shown, for example, in Figure 5, where the development of buyer 1's weighted average α is plotted in different experiments.

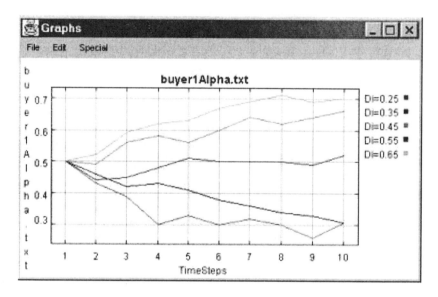

Fig. 5. The development of buyer 1's weighted average α in different experiments.

Buyers adapt the value they use for α and τ. As explained before, the weighted averages of α and τ for each agent indicate the emphasis they put on other agents' profitability vs. their trust in those other agents and on their own loyalty. The agents' learning can be represented as an adaptive walk across the fitness-landscape on the multi-dimensional problem space that is defined by α and τ. Such adaptive walks are illustrated in Figure 6.

This figure shows the combinations of weighted averages for α and τ on the x- and the y-axis, respectively, that each of the 12 buyers in experiment $d = 0.25$ maintains through time. A weighted average here is the sum of values for α and τ, multiplied by their selection probabilities, at each time step. Each agent starts at co-ordinates (w.a. for α and τ) = (0.5; 0.25); this is the average when the selection strengths for α (C_α) and τ (C_τ) are distributed evenly across the 5 possible values allowed – in $[0; 1]$ for α and in $[0; 0.5]$ for τ . The agents then follow different trajectories through the problem space. At the beginning, agents take relatively large steps through the problem space, in search for better performance. These large steps result from the fact that the selection strengths of α and τ are still more or less equal. Each agent's environment still changes radically with each time step. As agents form progressively better internal models of their environment, their behaviour becomes less erratic and the steps they take become smaller. Some local erratic movement remains from the fact that though small, the selection probabilities of other than currently successful evaluations are not zero, and some exploration still takes place.

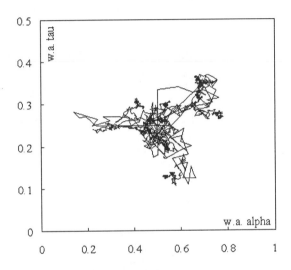

Fig. 6. The 12 buyers'adaptive learning in the space of α and τ in the first run, with differentiation $d = 0.25$.

The three different types of trajectory suggest that there may be locations in space that attract agents. One trajectory is a decrease of both loyalty and weight attached to trust (increasing α, decreasing τ), a second an increase in weight attached to trust (decreasing α), with little change in loyalty (stable τ), and the third entails a decreasing weight attached to trust and increasing loyalty (increasing α, increasing τ). So, both strategies of trust and loyalty and a strategy of opportunism can be learned to yield improved performance. This can make sense as follows: opportunistic agents engage in ongoing switching to seek out profitable partners, thereby relinquishing the advantage of learning by doing in ongoing relations, but gaining from the hunt for short term advantage, and perhaps gaining more benefits from economy of scale. Trust oriented agents go for learning by doing in more durable relations. One question which we still need to analyze is how this is realized in specific relationships: do partners with trust strategies team up, excluding opportunists who are then left to deal with each other? Does it occur that opportunists prey on trust-oriented agents who buyer have to be followed over all the time steps. The present implementation of the model theoretically allows for that, but this has not yet been added.

The process is accompanied with progressively higher levels of performance (profit), because realized profit drives the learning process. The system can be thought of as annealing over time, leading to a state in which all agents are attuned to one another. The extent to which such a state is robust under entry of other types of agents is the subject of future work.

We need to investigate whether the results in Figure 6 are specific for that particular run, or also apply to other runs. In Figure 7, the graphs for the individual buyers are averaged across buyers and those averages are displayed for each of the 25 runs. A great deal of information is lost in this averaging across agents. Because the trajectories of individual agents lead away from the starting point in different directions, as shown in Figure 6, averaging inevitably leads to values around the centre (note the difference in scaling along the axes between Figures 6 and 7). With these averages we still see movements in all directions, but with much more clustering near the centre, for the reason indicated. However, we still see that even averages lead to both higher and lower loyalty. Weight attached to profit relative to trust tends to increase more than to decrease.

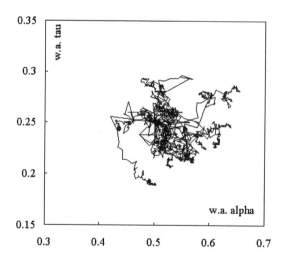

Fig. 7. The 12 buyers' adaptive learning in the space of α and τ, for all 25 runs with differentiation $d = 0.25$.

In sum, we find that developments may be as indicated by TCE, with decreasing loyalty and weight attached to trust. Here, agents learn to become opportunistic. However, there can also be increase of trust and trustworthiness, as we hypothesized. Now, we need to investigate more precisely when, in what 'worlds', the one occurs and when the other.

4.3   Different Cultures

In previous experiments, the plausibility of the model has been tested and results from the model have been analyzed concerning the two points of criticism of TCE. We are now ready to leave TCE behind and to start exploring the dynamics of the model in their own right. A final series of experiments has therefore explored the dynamics in the simulation when different 'cultures' are created, as a set-up to more elaborate testing of hypotheses concerning systems of innovation (Nooteboom 1999a).

These different cultures are defined in terms of the initialization of the strengths of the different values for α and τ, i.e. different sub-populations with differences in the importance attached to trust and loyalty. The left of Figure 8 shows how the strengths for the 5 possible values for a were initialized in the experiments reported above, and on the right the initialization in the some of the experiments with $d = 0.25$ reported below. In the previous experiments, the initial selection probabilities were distributed evenly over the 5 possible values for α, yielding absolute (relative) initial strengths of 0.15 (0.20). For τ, the same holds, except that 5 values between 0 and 0.5 were used, as explained above.

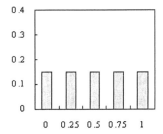 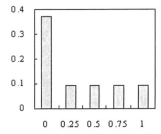

Fig. 8. Different ways of initializing strengths.

In the following experiments, the emphasis was either put on small or on large values for α and τ. Half of the total weight was assigned to one extreme value, and the other half was distributed evenly across the remaining 4 values, as illustrated in the right-hand picture in Figure 8. Rather than at (0.5; 0.25) as in the earlier experiment, the co-adaptive dynamics in the agents' problem spaces start at the four combinations of low and high weighted averages for α and τ, as shown in Figure 9.

On this level of double averages (across agents and runs) there is an indication of a prevalence of 'opportunistic' strategies, with decreasing weight to trust (increasing α) and some tendency towards decreasing loyalty (τ), for all starting points. How is this to be explained? There are basically two strategies for buyers. One is the scale strategy, where buyers seek a profit increase on the basis of economy of scale, by seeking a supplier who serves more than one buyer. This entails much switching and less emphasis on loyalty and trust. The other strategy is the learning by doing strategy, seeking an increase of profit in ongoing relations, on the basis of learning by doing. This entails less switching and more emphasis on loyalty and trust. The first strategy yields a faster increase in profit, since economy scale is instantaneous, while learning by doing takes time to build up. As a result, if the adaptation process starts from a zero level of learning by doing, the scale strategy leads to a higher value of α, since a high value attached to profit is rewarded by a fast increase of profit, which yields to a further increase of α. In other words, firms may get locked into the scale strategy. However, when a relation has lasted sufficiently long to generate significant learning by doing, the effect on α might switch. A high value attached to profit might then *prevent* buyers from switching, to preserve the profit from learning by doing. Then both buyers and suppliers may get locked into a learning by doing strategy. This would especially be the case if meanwhile high suppliers with multiple customers have reached their maximum number of customers, so that new buyers can no longer get in. This line of argument has several implications that can be tested against the evidence in figure 9.

1. At low levels of loyalty (τ), with more switching, the focus is on the scale strategy, which leads to higher values of α. This is confirmed in figure 9.

2. At higher levels of differentiation and specific investments (d), there is more reward for loyalty, in a learning by doing strategy, since that applies only to specific investments, so that τ tends to be higher. In figure 9, this is confirmed in three out of the four starting positions: upper right, lower right, and, to a lesser extent, lower left, but not upper left.

3. At higher levels of d, what would we expect to happen to a? The focus should be more on the learning by doing strategy, since a larger proportion of investments yields that opportunity. As indicated, the profit *increase* for learning by doing is slower than for a scale strategy. On the other hand, the *level* of the profit margin is higher, since more specific investments yield more differentiated profits. One argument now is, as argued above, that since profit increase is slower, α should reach lower levels, emphasising trust more than short term profit. However, once relations have lasted sufficiently long to generate high profit from learning by doing, higher levels of α might arise, to preserve that advantage. The evidence in figure 9 consistently points to the second effect: in all cases high levels of d lead to higher instead of lower levels of α.

Figure 9 also indicates a greater decline of τ when it starts from a high level than when it starts from a low level. In the fist case loyalty was apparently excessive. In the second case there is a suggestion of a minimum level of loyalty.

One conclusion is that it is not easy to interpret the results on the aggregate level of averages across buyers. In figure 9 even more information is lost than in figure 7. The 'process of becoming' at the level of individual agents needs to be studied in order to gain insight into why agents behave in the way they do.

5   Conclusions and Further Research

Some of the simulation results confirm TCE, and others confirm our objections
to it. In favour of TCE, the results confirm that higher product differentiation
and specific investments favour 'integration', i.e. 'make' over 'buy'. Also, there
is a tendency in profit-based adaptation to go towards less weight on trust and
less loyalty.

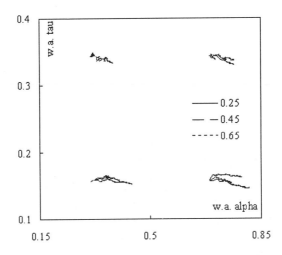

Fig. 9. Buyers' adaptive learning in the space of α and τ in different experiments,
starting from different initial positions.

However, counter to TCE our suspicions are confirmed that efficient outcomes
are often not reached due to the complexities and unpredictabilities of interaction. A new insight is that since higher product differentiation, together with
higher specific investments, leads to more 'make', and this yields less complexity of relations, efficient outcomes are more easily reached. Also, the tendency
towards lower weight to trust and less loyalty is visible only on the basis of
averages. Underlying those averages there are many cases of increasing weight
to trust and increasing loyalty. Furthermore, while on the whole loyalty tends
to decline, it does not go to zero. This suggests that perhaps agents must not
become too loyal, but some minimum loyalty is necessary.

In further experiments, we need to do two things. First, we need to analyze,
on the disaggregate level of individual agents, to what extent behind the averages
'trusting loyalists' seek each other out, and opportunists are left to deal with each
other. How often do opportunists prey on loyalists, does this happen only for
limited periods of time, or only when the loyalists have no options to switch?
When, if ever, does it occur that opportunists learn to become loyalists? Is it
ever viable to be a loyal 'sucker'?

Second, we need to investigate the viability of trust and loyalty under different values for other parameters, such as: the profit bonus of differentiated products, the relative strengths of effects of scale and learning by doing, the maximum number of buyers that a supplier can serve. Remember that learning by doing favours loyalty, and economy of scale favours switching to suppliers. Therefore the outcome is likely to be sensitive the relative benefits of scale versus learning by doing.

Another type of further research is to extend the model. In our view, the following extensions have priority:

1. In the present model, switching cost (cost of breaking a relation) derives exclusively from the loss of the benefits of cumulative learning in an ongoing relation. There is no switching cost due to the loss of specific investments, which forms a crucial part of TCE. This needs to be included, so that we can simulate different regimes concerning who pays for the loss: the supplier making the investment, both supplier and buyer as a result of shared ownership, or the principle that he who breaks pays. The inclusion of this type of switching cost complicates the model, since it requires assumptions concerning the capital intensity of profit (size of investment relative to profit margin), the period in which investments are amortized (economic life of investments), and linear or non-linear amortization. Also, one would need to build in the renewal of investments when relations last beyond the depreciation period. Furthermore, one may need to calculate preference scores and matchings in several rounds.

2. One justification for the assumption of efficient outcomes in TCE is evolutionary: inefficient outcomes are selected out, i.e. firms earning insufficient profits drop out. This needs to be brought in. That entails that the model is extended with exit and entry. Entry would also allow us to also investigate the robustness of evolved loyalty and trust under entry of opportunists. Does prevailing trust then break down, or do entrants adapt by learning to be loyal? Does the population perhaps split up in loyalists and opportunists? This is helps us to investigate the proposal by Hill (1990) that loyalist societies have competitive advantage, due to reduced transaction costs and more fruitful collaborative relations. That may be the case only if such a society closes itself off from new entry by possible opportunists. Under what conditions does existing trust unravel under such entry?

3. In the current model, the degree of product differentiation, which determines the extent to which investments are specific, thus causing switching costs, is exogenous. We consider the possibility of endogenizing it as another choice that is to made adaptively.

4. A third option we consider is to build in a reputation mechanism. Currently, trust is based on one's own experience with loyalty of partners. A reputation mechanism would entail that such experience is shared with others. This is likely to have significant effects on the development of trust and loyalty.

References

Albin, Peter S. and Duncan K. Foley, 1992, Decentralized, dispersed exchange without an auctioneer: A simulation study, Journal of Economic Behaviour and Organization 18(1), 27–51.

Alchian, Armen A, 1950, Uncertainty, evolution, and economic theory, Journal of Political Economy 58(3), 211–221.

Arthur, W. Brian, 1991, Designing economic agents that act like human agents: A behavioral approach to bounded rationality, American Economic Review 81(2), 353–359.

Arthur, W. Brian, 1993, On designing economic agents that behave like human agents, Journal of Evolutionary Economics 3(1), 1–22.

Arthur, W. Brian, John H. Holland, Blake LeBaron, Richard Palmer and Paul Tayler, 1997, Asset pricing under endogenous expectations in an artificial stock market, in: W. Brian Arthur, Steven N. Durlauf and David A. Lane, eds., The Economy as an Evolving Complex System, II, Santa Fe Institute Studies in the Sciences of Complexity Proceedings, Vol. XXVII (Addison-Wesley, Reading, MA) 15–44.

Birtwistle, Graham M., Ole-Johan Dahl, Bjørn Myhrhaug and Kristen Nygaard, 1973, SIMULA Begin, (Studentlitteratur, Lund, Sweden).

Booker, Lashon B., David E. Goldberg and John H. Holland, 1989, Classifier systems and genetic algorithms, Artificial Intelligence 40(1–3), 235–282.

Coase, Ronald H, 1998, The new institutional economics, American Economic Review 88(2), 72– 74.

Epstein, Joshua M. and Robert L. Axtell, 1996, Growing Artificial Societies: Social Science from the Bottom Up (Brookings Institution Press/The MIT Press, Washington, DC/Cambridge, MA).

Friedman, Milton (1953). The methodology of positive economics, in: Essays in Positive Economics (The University of Chicago Press. Chicago, IL), pp. 1–43.

Gale, David and Lloyd S. Shapley, 1962, College admissions and the stability of marriage, American Mathematical Monthly 69(January), 9–15.

Gambetta, Diego, ed., 1988, Trust: The Making and Breaking of Cooperative Relations (Basil Blackwell, Oxford).

Goldberg, David E., 1989, Genetic Algorithms in Search, Optimization and Machine Learning (Addison-Wesley, Reading, MA).

Gulati, Ranjay, 1995, Does familiarity breed trust? The implications of repeated ties for contractual choice in alliances, Academy of Management Journal 38(1), 85–112.

Hodgson, Geoffrey M., 1998, The Political Economy of Utopia: Why the Learning Economy Is Not the End of History (Routledge, London).

Holland, John H., 1992, Complex adaptive systems, Daedalus 121(1), 17–30.

Holland, John H. and John H. Miller, 1991, Artificial adaptive agents in economic theory, American Economic Review 81(2), 365–370.

Holland, John H., Keith J. Holyoak, Richard E. Nisbett and Paul R. Thagard, 1986, Induction: Processes of Inference, Learning, and Discovery (The MIT Press, Cambridge, MA).

Kirman, Alan P. and Nicolaas J. Vriend, 2000, Evolving market structure: An ACE model of price dispersion and loyalty, Journal of Economic Dynamics and Control, this issue.

Klos, Tomas B., 1999a, Decentralized interaction and co-adaptation in the repeated prisoner's dilemma, Computational and Mathematical Organization Theory 5(2), 147–165.

Klos, Tomas B., 1999b, Governance and matching, Research Report 99B41, Research Institute and Graduate School SOM.
http://www.ub.rug.nl/eldoc/som/b/99B41/.

Klos, Tomas B. and Bart Nooteboom, 2001, Agent based computational transaction cost economics, Journal of Economic Dynamics and Control, 25, 503–526.

Koopmans, Tjalling C., 1957, Three Essays on the State of Economic Science (McGraw Hill, New York).

Lane, David A., 1993, Artificial worlds and economics, part II, Journal of Evolutionary Economics 3(3), 177–197.

Lewicki, R.J. and B.B. Bunker., 1996, 'Developing and maintaining trust in work relationships', in R.M. Kramer and T.R. Tyler (eds), Trust in organizations: Frontiers of theory and research, Thousand Oaks: Sage Publications, 114-139.

McFadzean, David and Leigh S. Tesfatsion, 1999, A C++ platform for the evolution of trade networks, Computational Economics 14, 109–134.

Miller, John H., 1996, The coevolution of automata in the repeated prisoner's dilemma, Journal of Economic Behavior and Organization 29(1), 87–112.

Nooteboom, Bart, 1992, Towards a Dynamic Theory of Trans-actions, Journal of Evolutionary Economics, 2, 281–299.

Nooteboom, Bart, 1993, An analysis of specificity in transaction cost economics, Organization Studies 14(3), 443–451.

Nooteboom, Bart, 1996, Trust, Opportunism and Governance: a Process and Con-trol Model, Organization Studies, 17 (6), 985–1010.

Nooteboom, Bart, 1999, Inter-Firm Alliances: Analysis and Design (Routledge, London).

Nooteboom, Bart, 2002, Trust: forms, foundations, functions, failures and figures, Cheltenham UK: Edward Elgar, forthcoming.

Nooteboom, Bart, Hans Berger and Niels G. Noorderhaven, 1997, Effects of trust and governance on relational risk, Academy of Management Journal 40(2), 308–338.

Pagano, U., 1999, Veblen, new institutionalism and the diversity of economic institutions, Paper for the Conference of the European Association for Evolutionary Political Economy, Prague.

Péli, Gábor L. and Bart Nooteboom, 1997, Simulation of learning in supply partnerships, Computational and Mathematical Organization Theory 3(1), 43-66.

Pettit, Ph. (1995b), 'The virtual reality of homo economicus', The Monist, 78(3), 308-329.

Roth, Alvin E. and Marilda A. Oliveira Sotomayor, 1990, Two-sided Matching: A Study in Game-theoretic Modeling and Analysis, Econometric Society Monographs, Vol. 18 (Cambridge University Press, Cambridge (UK)).

Stanley, E. Ann, Dan Ashlock and Leigh S. Tesfatsion, 1994, Iterated prisoner's dilemma with choice and refusal of partners, in: Chris G. Langton, ed., Artificial Life III, Santa Fe Institute Studies in the Sciences of Complexity Proceedings, Vol. XVII (Addison-Wesley, Reading, MA) 131–175.

Tesfatsion, Leigh S., 1997, A trade network game with endogenous partner selection, in: Hans M. Amman, Berc Rustem and Andrew B. Whinston, eds., Computational Approaches to Economic Problems, Advances in Computational Economics, Vol. 6 (Kluwer, Dordrecht) 249– 269.

Tesfatsion, Leigh S., 2000, Structure, behavior, and market power in an evolutionary labor market with adaptive search, Journal of Economic Dynamics and Control, forthcoming.

Vriend, Nicolaas J., 1995, Self-organization of markets: An example of a computational approach, Computational Economics 8(3), 205–232.

Weisbuch, Gérard, Alan P. Kirman and Dorothea K. Herreiner, forthcoming, Market organisation and trading relationships, Economic Journal.

Williamson, Oliver E., 1975, Markets and Hierarchies: Analysis and Antitrust Implications (The Free Press, New York).

Williamson, Oliver E., 1985, The Economic Institutions of Capitalism: Firms, Markets, Relational Contracting (The Free Press, New York).

Williamson, Oliver E., 1993, Calculativeness, trust, and economic organization, Journal of Law and Economics 36(1), 453–486.

Winter, Sidney G., 1964, Economic "natural selection" and the theory of the firm, Yale Economic Essays 4(spring), 225–272.

Yelle, L.E., 1979, The learning curve: Historical review and comprehensive survey, Decision Sciences 10, 302–328.

Zand, Dale E., 1972, Trust and managerial problem solving, Administrative Science Quarterly 17(2), 227–239.

Zucker, Lynn G., 1986, Production of trust: Institutional sources of economic structure, 1840– 1920, in: B.A. Staw and L.L. Cummings, eds., Research in Organizational Behavior, Vol. 8 (JAI Press. Greenwich, Conn) 53–111.

Experiments in Building Experiential Trust in a Society of Objective-Trust Based Agents

Mark Witkowski, Alexander Artikis, and Jeremy Pitt

Intelligent and Interactive Systems Group
Department of Electrical and Electronic Engineering
Imperial College of Science, Technology and Medicine
Exhibition Road, London SW7 2BT
United Kingdom
{m.witkowski, a.artikis, j.pitt}@ic.ac.uk

Abstract. In this paper we develop a notion of "objective trust" for Software Agents, that is trust of, or between, Agents based on actual experiences between those Agents. Experiential objective trust allows Agents to make decisions about how to select other Agents when a choice has to be made. We define a mechanism for such an "objective Trust-Based Agent" (oTB-Agent), and present experimental results in a simulated trading environment based on an Intelligent Networks (IN) scenario. The trust one Agent places in another is dynamic, updated on the basis of each experience. We use this to investigate three questions related to trust in Multi-Agent Systems (MAS), first how trust affects the formation of trading partnerships, second, whether trust developed over a period can equate to "loyalty" and third whether a less than scrupulous Agent can exploit the individual nature of trust to its advantage.

1 Introduction

Software Agents are increasingly being required to make decisions and act locally, but also operate in the context of a "global" Multi-Agent Society (MAS). As these Agents become fully autonomous they become forced to make decisions about when and when not to engage (for instance to request information, to delegate important tasks or to trade) with other Agents. They must rely on internalised beliefs and knowledge about those other Agents in the society. This reliance on beliefs forms the basis of a *trust relationship* between intentional entities.

The trust relationship, in its broadest sense, has proved difficult to define [7], [8], [10], [15], [16], [17], [26]. We synthesise the following as a working definition, suited to the purposes of this paper. "*Trust is the assessment by which one individual, A, expects that another individual, B, will perform (or not perform) a given action on which its (A's) welfare depends, but over which it has restricted control*". Trust therefore implies a degree of dependency of A on B. This dependency may be reciprocal. Where the dependency relationship is asymmetric and one individual gains control over the other the relevance of the trust relationship

R. Falcone, M. Singh, and Y.-H. Tan (Eds.): Trust in Cyber-societies, LNAI 2246, pp. 111–132, 2001.
© Springer-Verlag Berlin Heidelberg 2001

is weakened for both A and B [16]. Equally, as the element of imposed compulsion in the relationship between individuals increases, the role of trust recedes. Similarly, the role of trust is reduced as the protagonists A and B acquire more complete information about each other (when they may accurately assess the future outcome of each transaction) [17]. Williams [26] summarises the trust relationship: *"agents co-operate when they engage in a joint venture for the outcome of which the actions of each are necessary, and where the necessary action by at least one of them is not under the immediate control of the other"*. The trust relationship may further be subject to exogenous events under the control of neither party, which may or may not affect the relationship [16].

Autonomous software Agents face all these issues, dependency on others, restricted control, incomplete information and the effects of exogenous events. It is little wonder, then, that the issues of trust between Agents should attract attention. Until an adequate system of compunction is widely adopted (through legislation, or by mutual agreement, for instance) this situation is likely to remain. Griffiths and Luck [11] emphasise the notion of trust as a reciprocal of risk, in the context of co-operative planning between Agents. Marsh [17] considers the risk/benefit relationship for Agents in a Distributed AI context. Castelfranchi and Falcone [7] divide the notion of trust estimation into component belief types that one Agent might hold with regard to another. They argue that such beliefs may be combined to form a *Degree of Trust* ($\mathbf{DoT}_{XY\tau}$) measure, which may in turn be used to decide whether a task of type (τ) should, or should not, be preferentially delegated by Agent X to some other Agent, Y. Jonker and Treur [15] present a formalised framework for the description of trust based on sequences of experiences between Agents.

The concept of trust within a society is closely allied to that of reputation ([3], [6], [28]). Reputation systems provide a mechanism by which individual Agents within their society can obtain information about other Agents without, or prior to, direct interaction and can lead to gains for the individuals and society as a whole [6]. We argue that trust should be based, whenever possible, on direct experience rather than on accumulated social attitude or shared reputation. As in real life, there is a limit to what can be achieved by wondering about what another entity might, or might not, do in any particular circumstance. We recognise that the definition of trust both as a function of accumulated beliefs and as a function of direct experience will be important to the construction of Agents and Agent Societies in the future. Such direct experiences can form an *objective trust measure* – the trustworthiness of another Agent put to the test and recorded as the basis of selecting that individual for future dealings. Such direct observation methods are important as they serve to ground in experience other assessed trust and reputation mechanisms.

In this paper we consider *objective Trust-Based Agents* (oTB-Agents), Agents that select who they will trade with primarily on the basis of a trust measure built on past experiences of trading with those individuals. The purpose of this work is to be able to investigate some important questions that arise when Agents are given a "free choice" as to whom they will co-operate with. This paper will consider three questions:

1. What happens when Agents who rank experiential trust and trustworthiness highly form into trading societies?
2. Does a trust relationship established between Agents over a period of time equate to loyalty between those Agents when trading conditions become difficult?
3. Trust, however it is evaluated, is personal; is it in an Agent's interest to appear trustworthy in some cases, and not care in others?

We take a practical and experimental approach in our investigations. To this end we present a scenario where Agents must choose who to trade with on the basis of a trust relationship developed between them and then adopt a concrete example within which to discuss and evaluate oTB-Agents. Section two provides an overview of that test domain. By describing the mechanism within this concrete example, we do not intend to convey any presumption that its application should be restricted to this or any particular application area, as we do not consider this to be the case. Section three defines the main functional components of an oTB-Agent. Section four briefly describes our Agent simulator and then presents the results of some experiments that shed light on the questions just posed. Finally, we discuss related and future work and draw some conclusions.

2   The Trading Scenario

In order to test this notion of objective trust we establish a simulated trading environment in which many individual Agents must select partners with which they will trade on an ongoing basis. This continued trading within a closed community allows trust relationships to be made, sustained or broken over an extended period. The trust relationship must be essentially symmetrical; each party must be able to behave in a trustworthy or untrustworthy way towards others, and have others behave similarly towards them. To complete these experiments the individual Agents must also be subject to various exogenous events (those beyond their control), which force them to act in an untrustworthy way towards certain trading partners. We adopt a specific example, which is described next.

Fig. 1 shows an idealised model for a telecommunications *Intelligent Network* (IN). The IN provides an infrastructure in which different types of Agent may form a trading community, as well as acting as an interface layer between end-user consumers of a communications service and the underlying telecommunications network which will transport voice and data information between geographically distinct points.

We consider two distinct Agent types in this paper. *Service Control Point (SCP) Agents* are associated with *Service Control Points*, access portals to the telecommunications network. *Service Switching Point (SSP) Agents* serve *Service Switching Points*, providing access points for consumers of telecommunications services. There may be a large number of SCP and SSP Agents forming a single

IN. Each SCP acts as an agent or broker for the suppliers of telecommunications bandwidth and is tasked with ensuring that the available bandwidth is sold. Conversely, each SSP acts as agent or broker for end-consumers of telecommunications services, tasked with ensuring that sufficient bandwidth is reserved to meet the needs of those consumers.

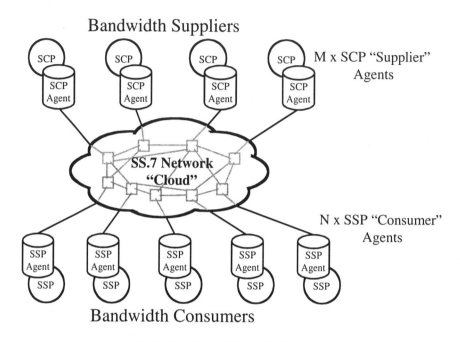

Fig. 1. Intelligent Networks Scenario.

In this model message passing between SCP and SSP Agents is assumed to take place over an SS.7 network and to use a contemporary Agent Communication Language and Protocol, such as FIPA-ACL [9], [13]). Beyond requiring that the transmission of messages between Agents is timely and reliable, we will not consider related issues of inter-Agent communication further in this paper.

Agent Architecture models for IN management have proved to be rich ones for investigating issues directly related to resource allocation and load control in the context of current telecommunications systems ([13], [19], [20], [23]). Rather than concerning ourselves with issues relating to overall performance of the network, we will concentrate on the effects of trading decisions based on "objective Trust Based" (oTB) principles. We will focus on the performance of individual Agents from the perspective of the degree to which they trust, and are trusted by, other Agents in the society. In maintaining this focus on issues relating to trust we have developed a "trading scenario", which gives both SCP and SSP Agents the opportunity to behave in a trustworthy or untrustworthy way in their dealings with fellow Agents. This then forms a basis on which individual Agents select the Agents they will trade with in future.

2.1   The Trading Cycle

Trading is divided into equal time slots, called a *trading cycle*. At the beginning of each trading cycle every SSP (customer) Agent receives a demand for resource (bandwidth in the scenario) and makes bids to SCP (supplier) Agents to cover that demand. SSP Agents must select SCPs they trust to offer them the resource they require. If the SCP does not offer to cover an SSP's bid for bandwidth resource, then the SSP has reason to regard that SCP as untrustworthy. While demand may vary between trading cycles, the total amount of resource available is taken as fixed. Each SCP Agent must attempt to distribute its supply of resource to SSP Agents that it trusts to pass that resource on to its end-users. Any resource not taken up by SSP Agents is deemed lost, to the detriment of the SCP Agent. SSP Agents that fail to use resource offered to them are considered untrustworthy.

All SSP and SCP Agents each maintain a *trust vector*, recording the opinion the Agent holds about the trustworthiness of each of the other Agents with which it can trade. The trust vector forms the primary source for selecting trading partners, and is itself updated after each transaction.

Each trading cycle involves three transaction steps (each corresponding to an ACL performative between individual Agents). First, the *bid step*, in which SSP Agents receive their demand load and issue bids to SCP Agents to meet that load. Second, the *offer step*, in which SCP Agents make offers of resource in response to bids they receive. Third, the *utilisation step*, in which SSP Agents distribute the resource units they have been offered to their customers, and notify the SCP that offered the resource whether or not they utilised all the allocation they were offered.

2.2   The Bid Step

At each trading cycle every SSP Agent receives a quantity of demand from its customer base, which is the sum of their (the customers') estimates of the resource they require for the next trading cycle. Each SSP must then select one or more SCP Agents it trusts using an *allocator function*, and issue a *bid message* performative to them indicating the number of units of resource it requires. An SSP Agent may dishonestly (or perhaps prudently) *overbid* its requirement, thereby ensuring it will receive at least as much resource as it requires. In doing so it risks having to return unused units, and be seen as untrustworthy by the SCP Agent that reserved resource for it.

2.3   The Offer Step

Each SCP Agent receives a quantity of units bid from SSPs willing to trade with it. SCP Agents select which SSP bids it wishes to honour using a *quantifer function*, the choice being derived from the Agent's trust vector. The SCP Agents then communicate the offers of resource they are prepared to make back to the SSPs that made the original bids, the *offer message* performative. An SCP may not offer, in total, more resource units than it has access to. To do so would, in this scenario, introduce another round of transactions.

2.4   The Utilisation Step

Once an SSP Agent has received all the offer messages from SCP Agents, it will attempt to satisfy the customer demand for the current trading cycle from the offers of resource allocation that it has obtained. If it has received more resource than it requires it returns the excess to one or more SCP Agents on the basis of a *utilisation function*. Returns are notified to SCP Agents in a *utilisation message* performative. Also at this step the SSP Agent updates it trust vector using its SSP trust function, on the basis of the difference between the quantity the SSP Agent bid for against the quantity it received from SCP Agents. We assume accountability, in that an SCP Agent can meter units actually consumed at the request of an SSP Agent, so that an SSP cannot just request an unlimited number of units and just discard the excess (thereby appearing trustworthy to the SCP). On the other hand, each SSP is free to return unused units to any SCP, thereby managing its trust relationships.

Finally, on receipt of the utilisation messages, each SCP Agent can update its own trust vector according to its SCP *trust function* by comparing the quantity of resource requested against that actually utilised.

We treat the resource (bandwidth) as a true commodity. Any SSP may request resource from any SCP. We further treat the resource to be a fixed price item. Agents may not "spend more" to secure extra supplies in times of shortage, or reduce their prices in times of oversupply. When supply and demand are mismatched individual Agents must decide which Agents they will favour over others, this is at the heart of the "does trust beget loyalty?" question posed earlier.

There is no overall control or centralised mediation in this system model (as, for instance, in the auction model of Patel, *et al.*, [18]). Each Agent makes its trading decisions based on its past experiences of trading with other Agents in its community, updating its trust vector, and so affecting its future decisions, based on each new transaction. In the model, Agents that do not adhere to the communications and transaction protocols are excluded from the trading arrangement. Messages sent inappropriately, such as an SCP offer where no bid was made, can be discarded and the sender considered "untrustworthy" for attempting to supply an unsolicited service.

3   The Allocator, Quantifier, Utilisation and Trust Functions

This section describes the SSP-Allocator, SCP-Quantifier, SSP-Utilisation and the Trust functions used by SCP and SSP Agents in detail. Together these five functions encapsulate the key components of oTB-Agents. Fig. 2 illustrates the internal structure of the SSP and SCP trading Agents used in the experiments to be described later, and indicates the order in which each of the five functions is invoked in the context of the overall trading cycle described in the previous section.

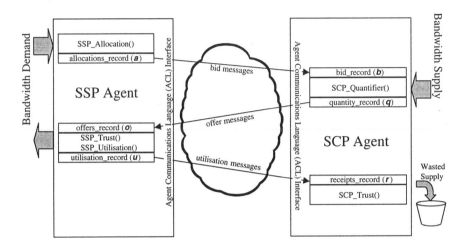

Fig. 2. The Trading Scenario.

In a society of N SCP Agents trading with M SSP Agents, the trust vector owned by the nth SCP Agent will be represented by $_n t$, its trust rating of the mth SSP, a scalar value, by $_n t_m$. Conversely, the mth SSP's $(_m t)$ trust rating of the nth SCP Agent by $_m t_n$. Individual trust ratings are scaled from 0 (complete distrust) to 1.0 (complete trust). The use and management of these trust values is central to the operation of an oTB-Agent, they are the principal way in which other Agents are selected to trade with. The manner in which it is used, and the mechanism by which it is updated, define important aspects of an Agents apparent "personality" (the way it appears to other Agents) within the society. The *allocations record* (a), *offer record* (o) and *utilisation record* (u) are message buffers used by SSP Agents to prepare messages for sending (a and u), or receiving (o) messages from SCP Agents. The *bid record* (b, receive), *quantity record* (q, send) and *receipts record* (r, receive) are used by SCP Agents to buffer messages to and from SSP Agents. They employ the same indexing notation as t.

3.1   SSP Allocator Function

The SSP allocator function divides the total demand (`actual_demand`) received by an SSP Agent for the current trading cycle into smaller units and populates an *allocations record*, $_m\mathcal{A}_n$, which holds the number of units of resource the SSP Agent m will be requesting from SCP Agent n.

The allocator function is controlled by three parameters. (1) The *overbid rate*, obrate, which determines how much extra resource the Agent will bid for above its actual demand. Overbid is expressed as a percentage. (2) *The split rate*, srate, which determines how many SCP Agents will receive bids from this SSP Agent. This effectively ameliorates the risk for the SSP Agent that any particular SCP Agent will refuse it supply. The split rate is expressed as an integer ≥ 1, but $\leq$ number of SCP Agents. (3) The *exploration rate*, erate, which determines the probability with which the Agent will ignore its trust ratings and send a bid to a random SCP Agent, where 0 represents no exploration of the market and 1.0 causes the SSP Agent to always select suppliers at random. The exploration rate parameter addresses a practical problem familiar in the reinforcement learning paradigm, that of balancing the advantages to be gained from trading with known and already trusted partners with the opportunity to discover better partners from the larger pool ([21]).

The allocator function is best described procedurally:

For each SSP Agent m do:

1. Clear $_m\mathcal{A}$
2. Set demand $\leftarrow$ actual_demand * obrate
3. Set bid_packet_size $\leftarrow$ demand / srate
4. If (rand < erate) Set $_m\mathcal{A}_r \leftarrow$ bid_packet_size
     where $_r$ is a randomly selected SCP Agent and
     rand is a randomly generated number, 0 .. 1.0
5. Else for SCP Agent x, where x is $\max(_m\mathcal{t}_x)$ and $_m\mathcal{A}_x = 0$
     Set $_m\mathcal{A}_x \leftarrow$ bid_packet_size
6. Repeat from step 4 until all bid packets allocated.

Step 5 successively selects the most trusted, then the next most trusted until all the bid packets have been allocated. Once the allocator procedure is completed the SSP Agent issues a bid message to every SCP Agent where $_m\mathcal{A}_n > 0$ (i.e. a bid has been allocated). Apart from the random selections, bids have been sent to the most trusted trading partners.

3.2   SCP Quantifier Function

The SCP quantifier function distributes the SCP Agent's limited supply amongst all those SSP Agents that made bids, it does so on the basis of trust, as recorded in its trust vector. The function is unparameterised. Received bids are recorded in the *bid record* $_n\mathcal{b}$, the SCP quantifier function populates the *quantity record* $_n\mathcal{q}$, which records the offers to be made. If the total of bids (total_bid_value) received by the Agent total less than the available supply, the value of each bid is simply transferred to the quantity record, as all SSP Agent bids can be satisfied. When bids exceed supply the following procedure is invoked to distribute the available supply on the basis of trust:

For each SCP Agent n do:
  While `total_bid_value` > 0

   For SSP Agent x, where x is $\max(_n t_x)$ and $_n q_x = 0$
   Set $_n q_x \leftarrow {}_n b_x$ if `total_bid_value` $\geq {}_n b_x$
    else $_n q_x \leftarrow$ `total_bid_value`
   `total_bid_value` $\leftarrow$ `total_bid_value` - $_n q_x$

Offer messages are issued from $_n q$ notifying the bidding SSP Agents whether their bid has been successful or not, SSP Agents note these offers in their *offers record*, $_m o$. This procedure effectively assigns to the SSP Agents that an SCP Agent trusts the most all the supply they want, giving priority to the most trusted Agents first, until all the supply is used up. The remaining Agents are rejected. Other quantification strategies can be implemented, for example equable distribution where each bidder receives a fair share of the supply, but these are not considered further here.

3.3   SSP Utilisation Function

When an SSP Agent has bid for, and received, more units than it actually requires it may return these excess units to unfortunate SCP Agents, who have lost the opportunity to use them and the units are wasted. If demand exceeds offers, the SSP Agent satisfies its customers as best it can, and transfers all the used offers from $_m o$ to its *utilisation record*, $_m u$ (full utilisation). When offers exceed demand, $_m u$ is populated thus:

For each SSP Agent m do:
While `total_offer_value` > 0
For SCP Agent x, where x is $\max(_m t_x)$ and $_m u_x = 0$
Set $_m u_x \leftarrow {}_m o_x$ if `total_offer_value` $\geq {}_m o_x$
  else $_m u_x \leftarrow$ `total_offer_value`
`total_offer_value` $\leftarrow$ `total_offer_value` - $_m u_x$

The SSP Agents utilises offers from SCP Agents with which it has the best trust relationships preferentially, and risks damaging relationships that are already weaker. Entries in $_m u$ are transmitted to SCP Agents who made offers as *utilisation messages*, and recorded by the receiving SCP Agent in its *receipts record*, $_n r$. The SSP Agent suffers no actual penalty, except the loss of credibility with its supplier, for returning offers unused.

3.4   SSP Trust Function

An SSP Agent's trust vector is updated on the basis of the perceived reliability of SCP Agents. This is determined on the basis of whether, or not, an SCP Agent honoured individual bids, $_m a_n$, with corresponding offers, $_m o_n$. A trust function takes two parameters, $\alpha (0 \leq \alpha \leq 1)$, the degree to which a positive experience enhances a trust vector element, and β $(0 \leq \beta \leq 1)$, the degree to which a negative experience damages the relationship. An individual SSP Agent trust vector element, $_m t_n$, is updated thus:

$_m t_n \leftarrow {_m t_n} - (\beta * {_m t_n})$, if a bid $_m a_n$ was issued, but no offer $_m o_n$ received, or

$_m t_n \leftarrow {_m t_n} + (\alpha * (1 - {_m t_n}))$, if offer $_m o_n \geq$ bid $_m a_n$ was issued, or

$_m t_n \leftarrow {_m t_n} + ((\alpha * ({_m a_n} - {_m o_n})) * (1 - {_m t_n}))$, if $_m o_n < {_m a_n}$, or

$_m t_n$ is left unchanged otherwise.

These formulations are normalised such that a string of positive experiences asymptotically moves $_m t_n$ towards 1.0, and a string of negative experiences moves it towards 0.0. The function matches our intuition that trust is most enhanced by getting exactly what we requested, partially enhanced by getting some of our request and damaged by being excluded. The formulation also conforms to our expectation that recent experiences are given greater weight that earlier ones, the effect of past events are increasingly discounted with each new experience, but never completely lost. Agents that adopt high values for α are generally more susceptible to single positive experiences, those that adopt a high β value more influenced by negative experiences.

3.5   SCP Trust Function

The SCP trust function is analogous to the SSP trust function, except that it is driven from a comparison of the resource offered, $_n q_m$, against that utilized, $_n r_m$.

$_n t_m \leftarrow {_n t_m} - (\beta * {_n t_m})$, if an offer $_n q_m$ was made, but no utilisation $_n r_m$ was made, or

$_n t_m \leftarrow {_n t_m} + (\alpha * (1 - {_n t_m}))$, if utilisation $_n r_m =$ offer $_n q_m$, or

$_n t_m \leftarrow {_n t_m} + ((\alpha * ({_n q_m} - {_n r_m})) * (1 - {_n t_m}))$, if $_n r_m < {_n q_m}$, or

$_n t_m$ is left unchanged otherwise.

4   Experiments in oTB-Agent Based Trading

We have prepared a simulator in order to investigate the properties of oTB-Agents in the trading situation described previously. The simulation is detailed in that it performs each step in the oTB-Agent algorithm for every Agent at each trading cycle, and emulates every communication message between Agents. The simulator allows the investigator to specify the number of SCP and SSP Agents that will participate, and to set the important parameters for both types of Agent, the α and β trust modification rates; and the overbid rate, the split rate and exploration rate (for SSP Agents). The investigator may single step the simulation, or run it for a pre-determined number of trading cycles, modify parameters and continue. The simulation provides a graphical indication of messages between Agents, and indicates the utilisation of bandwidth resource due to that Agent (as a percentage of the total possible). The investigator may also inspect the trust relationships between any single Agent and its trading partners. At the end of a simulation session logging files may be produced giving a complete record of the development of the trust vectors.

4.1 Experiment One

Experiment one will investigate the effects of load on the relationship between SSP and SCP Agents. We establish trading communities of 10 Suppliers (SCP) and 20 Consumer (SSP) Agents. All SCP Agents are the same ($\alpha = \beta = 0.25$), as are all SSP Agents ($\alpha = \beta = 0.25$, srate $= 4$, erate $= 0.2$, obrate $= 0\%$ (i.e. no overbid)). In these experiments all suppliers receive an identical allocation of bandwidth, and all consumers have an equal demand placed on them. All SCP and SSP Agents are identical and treated identically to ensure that the effects of the oTB-Agent procedure are placed in a "fair" trading situation (our first question from section 1).

This experimental investigation is in three parts, and the results are summarised in figs. 3, 4 and 5. In each part the supply of bandwidth resource is successively restricted in relation to demand, to cause an "overload". Under these circumstances SSPs must develop strong relationships in order to ensure supply (in the converse situation, SCPs are under pressure). Three separate runs are made, one where supply exactly matches demand (100% supply, fig. 3), one where supply is 75% of demand (125% overload, fig. 4) and one where supply is only 50% of demand (150% overload, fig. 5). Each graph in these figures indicates the changing trust relationships of a single Agent (SSP above, SCP below) to all its trading partners. In each case, the SSP graph (top) is one of 20, and the SCP graph (below) one of ten.

Each figure also highlights the relationship between specific pairs of Agents (fig. 3 between SSP #5 and SCP #3, fig. 4 between SSP #5 and SCP #8, and fig. 5 between SSP #9 and SCP #3). Note that each run is completely separate, starting with a new random initialisation, therefore Agent numbering in each figure is independent. At the start of each experimental run of 200 trading cycles every trust vector element in all Agents is seeded with an initial value random value in the range 0.499999 and 0.500001. In general, oTB-Agents do not have any "opinion" about the trustworthiness of other Agents (i.e. a trust value of 0.5) at the start of a trading session. This small random perturbation pre-disposes them to start trading with some Agents in preference to others. oTB-Agents are therefore initially *trust neutral*, [15], prior to gaining experience through trading.

In all instances we see that SSP and SCP Agents tend to "pair-off" very quickly. In the 100% supply case (fig. 3) we can see that SSP Agent builds trust relationships with SCP Agents #0, 2 and 7 quickly, followed by Agent #9 (highlighted with a '+') soon after. These are its preferential trading partners,

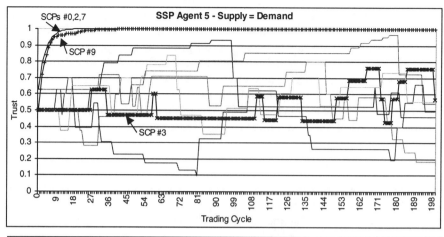

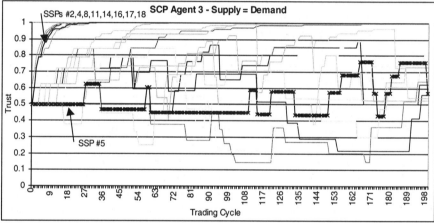

Fig. 3. Trust Relationships Between Agents with Balanced Supply and Demand.

but it partially trusts many other SCP Agents and trades with them from time to time (this occasional trading between SSP #5 and SCP #3 is highlighted with 'x' markers).

In the 125% loading case (fig. 4) we note that this "pairing-off" is more pronounced. Moreover, the number of preferred trading partners has dropped. This indicates that suppliers (who have the upper hand in this situation) prefer to maintain a smaller number of trusted customers, and serve them fully. In turn the customers must continue to bid to these suppliers regularly in order to safeguard their supply of bandwidth. The preferential trading partnership between SSP Agent #5 and SCP Agent #8 is highlighted in the middle row. This effect becomes ever more pronounced as supply is further restricted. At 150% overload (fig. 5) the gulf between those Agents that can trade because they succeeded in establishing a trust partnership and those that did not is

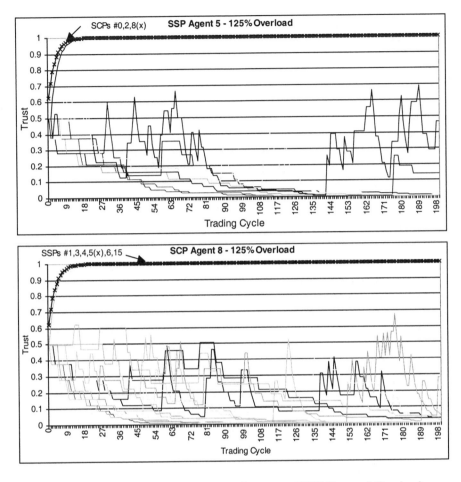

Fig. 4. Trust Relationships Between Agents at 125% Demand Overload.

very clear. SSP Agent #9 only secured one trust relationship (with SCP #3, highlighted '+'), and is clearly going to struggle for supply.

Fig. 6 makes explicit the overall relationship between the degree of trading trust an SSP Agent has been able to secure and its ability to deliver bandwidth to its customers. Each marker in the graph shows the average trust rating for each SSP Agent across all the SCP Agents, against its success in meeting demand. When supply equals demand (100% supply, diamond markers), the overall ability of an SSP Agent to deliver is hardly affected by its perceived trust rating (delivery rate is largely unaffected by overall trust rating). As supply is restricted, (125% overload markers), there is a clear correlation between trust rating and ability to deliver has developed. When supply is further restricted to 50% of demand (150% overload, triangle markers) the correlation is pronounced. The performance of each of the three sub-groups shown circled is directly pro-

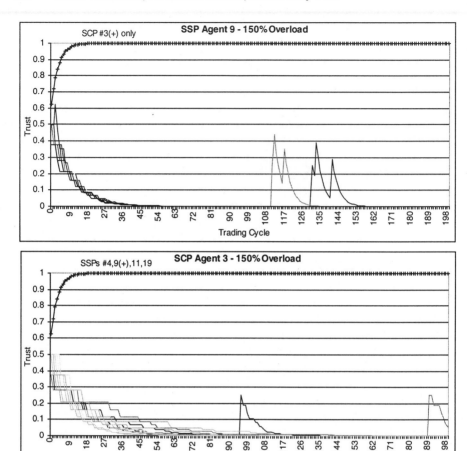

Fig. 5. Trust Relationships Between Agents at 150% Demand Overload.

portional to the number of suppliers with which the SSP Agent has managed to build a trading relationship. The worst performing group (group 3) only established a partnership with one other SCP Agent (SSP Agents #2, 3, 4, 9 and 13, with an average trust rating of 0.1 and a 0.252% delivery record). The higher group (group 1) comprises Agents #5, 11, 14, 16 and 18, with an average delivery record of 0.729%, established relationships with three suppliers. In this instance no SSP Agent formed a group with four partners under these conditions.

4.2   Experiment Two: The Effects of Changing Circumstances

Experiment two addresses our second question, as to whether establishing a trust relationship over a period of time will equate to loyalty when trading becomes difficult. We repeat the conditions of part one of experiment one (supply =

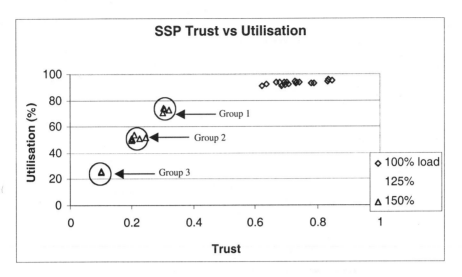

Fig. 6. The Effect of Trust Rating on SSP Delivery Performance

demand), except that at trading cycle 100 supply is reduced to 75% of demand (125% overload). Fig. 7 shows a pair of trust graphs linking the effect on the trust, and hence trade, relationship between SSP Agent #18 and SCP Agent #9 (highlighted 'x').

It is clear from inspection of these graphs (and the others in the set, not shown), that in addition to the loss of weaker trust relationships (as was the case in experiment one), suppliers have a marked tendency to discard their strong partners on a last-in first-out basis. It appears that, at least in this case, trust does give rise to loyalty.

4.3   Experiment Three: The Effect on Trust of "Greedy" Behaviour

To address our third question, whether an Agent can exploit the "personal" nature of the trust relationship, we perform an experiment in which the SSP Agents are divided equally into two groups. In one group (the "normal" group) they trade "honestly", only bidding for the units of bandwidth they actually require. The second group act "greedily", bidding for 150% of the units they require (obrate = +50). Supply is set to equal demand, and the other conditions are as before. Fig. 8 summarises the results obtained from running of this experiment.

The effects of this greedy behaviour are clear. While the average trust rating by all SCP Agents (front rank) of the greedy SSP Agents is far lower than that for the normal ones (0.391 vs. 0.680), their overall delivery performance (rear rank) is somewhat better (96.1% vs. 88.5%). They perform better because they receive more offers of bandwidth due to overbidding. The effect of the oTB-Agent procedure is to always preferentially buy from your preferred suppliers.

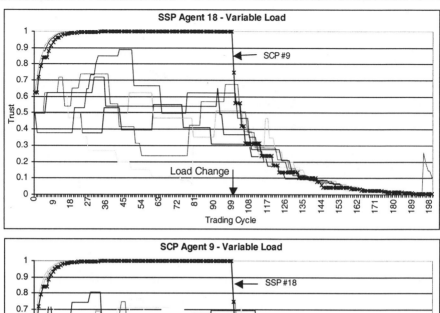

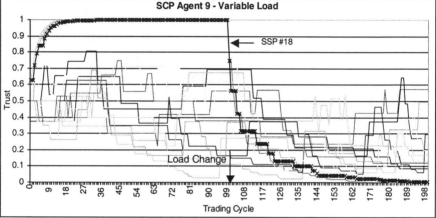

Fig. 7. The effect of Increased Load

So where the normal Agents have good relationships with their preferred suppliers, and reasonable relationships with others, greedy Agents have equally good relationships with their preferred suppliers, but very poor relationships with all the others, who they have treated badly. An element of duplicity, it seems, is still effective in a society where trust is otherwise highly valued.

5   Related Work

There exist a number of issues in the research for security and trust in MAS. We consider three in the context of this work, the role of enforced security measures relative to the social approach, the role of explicit vs. implicit cognitive modelling and the effects of centralised vs. decentralised control in Agent Societies.

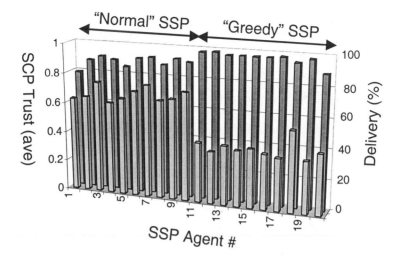

Fig. 8. The Effects of Overbidding

5.1 Cryptography and Network Security Techniques vs. Social Approaches

Wong and Sycara, [27] address a number of security and trust issues faced by MAS and provide an infrastructure to deal with such issues. They make use of techniques that are well known in the network security literature, and apply these techniques to MAS. They propose no measures relating to trust or honesty, with no way of ensuring that an Agent will carry out a task as expected, or of guiding an Agent to interact with other Agents that will *probably* be honest. Other approaches of dealing with security issues in MAS have been made by Thirunavukkarasu, Finin and Mayfield, [24] and He, Sycara and Su [12]. These approaches introduced, among a number of things, a number of new KQML performatives enabling Agents to interact in a secure manner. These researchers use classic network security techniques and do not propose any trust models.

It is obvious from this kind of research that network security alone is not sufficient, considering the requirements of multi-agent systems. Following a social approach to security in MAS, Biswas, Debnath and Sen [4] have proposed a model where Agents have relatively complex behaviours. They use a probabilistic mechanism in which an Agent A will decide whether or not to honour a request for help by Agent B. This mechanism takes into consideration previous observations of Agent B, as well as the additional costs incurred by Agent A from Agent B. These researchers demonstrate that Agents that adapt their trust models over time and use probabilistic decision mechanisms are able to successfully withstand the invasion of selfish and exploitative Agents.

oTB-Agents use similar mechanisms for adaptation and decision making. However, unlike oTB-Agents, the Agents of the model of Biswas, Debnath and Sen consider all of their previous observations equally before delegating a task.

oTB-Agents, using the trust function described in section 3, place extra weight on recent experiences, although they are influenced by all experiences between the two Agents.

5.2 Implicit vs. Explicit Cognitive Approaches

In implicit approaches, Agents use assessed probabilities to model the trustworthiness of the others. Schillo and Funk [22] conducted a number of simulations where Agents interact with each other using a modification of the prisoner's dilemma (i.e. the disclosed prisoner's dilemma with partner selection). Each single Agent builds a model of trustworthiness of the other Agents by gathering data on past behaviour and evaluating averages. When Agents are asked about their knowledge on other Agents, they are free to lie about their observations. Nevertheless, Schillo and Funk show that by averaging the values of a sufficient number of observations Agents can learn models almost twice as fast as other Agents that use only their own observations, while still reaching the same or better accuracy.

Schillo and Funk's Agents are characterised along two dimensions, being honest/dishonest and altruistic/egoistic. As with [4] the age of an observation is not taken into account. Furthermore no consideration is given as to the reliability of the sources that provide information about Agents.

Castelfranchi, Conte and Paolucci [6] have stressed the importance of reputation in relation to the modelling of trust. These researchers performed a set of experiments in order to simulate the role of reputation in the re-distribution of the costs of norm compliance in agent societies that included normative and non-normative (*cheater*) agents. They showed that communicating knowledge about others' behaviour leads to improved performance of the normative agents. It is important to note that, in contrast to Schillo and Funk's setting, in Castelfranchi, Conte and Paolucci's experiments, the communicating agents did not lie about their observations.

oTB-Agents do not exchange information about their past observations. Clearly, if oTB-Agents communicated their observations (as in the case of Schillo and Funk and Castelfranchi *et al.*) then the 'greedy' Agents of the third experiment would not perform as well as they did. Another major difference between the oTB-Agents and the work that was presented in this section is that, unlike oTB-Agents, the *recency* of an observation is not taken into account.

Explicit cognitive approaches (e.g. [3], [14]) appear more sophisticated, as they attempt to model the "mind" of the other Agents. Castelfranchi and Falcone [7] give a number of guidelines that should be taken into consideration when modelling the trustworthiness of other Agents. These authors separate the concept of trust from that of delegation and mention a number of beliefs that should exist before delegating a task to another Agent (i.e. competence, disposition, dependence beliefs, etc.) Jones and Firozabadi [14], use tools from modal logic to characterise aspects of the reasoning of the Agent who trusts the reliability of the information communicated to it.

5.3   Centralised vs. Decentralised Control over the Groups of Agents

In organisations where there is a form of centralised control, trust can be viewed as a three party relationship [7]. Agents trust the ability of the authority to assess contract violations and to punish the violators. Agents also trust that other Agents will not violate contracts because they respect/fear the authority. On the other hand, there exist groups of Agents with no form of centralised control. In these groups, Agents need to develop their social skills in order to avoid being exploited by deceitful Agents. The oTB-Agents described here exist in an environment without centralised control.

6   Discussion

The experiments show that oTB-Agents tend to form strong, tight, clusters of trading partners very quickly, and that these partnerships become increasingly important as supply and demand for the traded commodity becomes mismatched. Trust builds trust, but unreliability breeds indifference, *"trust is a peculiar resource which is increased, rather than depleted, through use"* [10]. The Agents modelled here show a clear preference for building strong relationships with trusted partners, sustaining successful partnerships and discarding less trusted partners when conditions turn unfavourable. "Deceitful" Agents, those who generally behave in an untrustworthy manner can still thrive in this community, as long as they maintain good trust relationships with a few key partners.

Reciprocal behaviours in a variety of forms are recognised as effective strategies for forming stable groupings with in larger community [4]. The oTB-Agents defined here appear to adopt an extended "tit-for-tat" attitude, as might be encountered in various game theoretic approaches, such as the Iterated Prisoner's Dilemma ([2], [22], [26]). The scenario presented here differs from the well-understood iterated prisoner's dilemma in that the selection is made on the basis of every transaction between the Agents. Simple Tit-for-tat strategies are considered to be insufficient for most domains of practical interest [4], these problems are largely overcome when the complete transaction history is considered. In addition, SSP oTB-Agents partially base their partner choice on the basis of exploration.

Variations of the formulation used here to evaluate trust find wide application in some of the more numerically orientated approaches to machine learning (such as reinforcement learning, [25], for example) and is ubiquitous, though by no means universal, in theories of natural learning. Jonker and Treur [15] propose a similar formulation for the "quantitative" component of their formalisation of trust. Despite its apparent simplicity the application of this formulation invariably imparts interesting behaviours to the systems that incorporate it.

It is clear from the experiments that a successful first transaction is central to establishing the inter-Agent trading relationship, and an area where the prior assessment of possible partners is critically important. Equally, were this trading community to be augmented with a "reputation" mechanism (such as those of [4], [22] and [28]), by which Agents entering the market could consult existing traders, then the "greedy" Agents of experiment three would be put at a disadvantage. Similar results, i.e. putting "greedy" Agents at a disadvantage, would be obtained in the case where members of the trading community had the ability to *observe* the behaviour and the interactions of the other Agents. In such a case, Agents would have an additional source of information that would enhance their decision-making process concerning their potential trading partners.

7   Current and Future Work

Our aim is to produce a formal specification of the trust model of oTB-Agents that encompasses all three components of a trust based trading relationship, reputation, belief based trust and objective (direct experience) trust. Each, we believe, has an important role to play at different times in the overall life of a trading partnership.

There exist a number of attempts to formalise the concept of trust in MAS ([8], [14], [15]). According to some of these approaches, the formal specification of trust models should include, among other things, decisions such as the use and formation of a trust evolution or update function as well as the properties that should hold for that function [15].

In order to be able to claim that our model of trust is widely applicable, a number of issues are to be addressed in the future. One such issue is the exchanging of observations about other Agents. Consulting other Agents has proved to be helpful in many experiments ([4], [22]). Nevertheless, in the case where a kind of reputation mechanism is used, then the reliability of the Agents information sources should be taken into consideration. Experiential evidence, such as that obtained by oTB-Agents can provide the foundation on which these mechanisms may be built. However it has been shown that the question of what information an agent should accept is a non-trivial task [5]. Several ways have been proposed to order incoming information and consequently decide what information to accept or reject [5]. As Agent behaviour becomes more complex, (e.g. having more opportunities to cheat during interactions), so the modelling of the trustworthiness of other Agents becomes increasingly complex. We intend to integrate our trust modeling with belief analysis and revision [7]. An Agent should be able to evaluate the competence, the willingness and the trustworthiness of another Agent before delegating a task. New parameters should be introduced in the trust modeling process, such as a risk threshold (i.e. how much is an Agent willing to risk the delegation of a task).

Our oTB-Agents trade and develop trust along a single dimension only. More sophisticated Agents will engage with other Agents for a variety of different reasons, and trust should be, in part, a function of the task being performed (Agent X may be reliable when performing task1, but unreliable on task2). We would therefore expect an Agent to maintain an estimate of trust about each task under these circumstances.

A number of other issues are worth considering in future experiments and simulations of Agent communities, such as silent communication, feelings and affective trust. Having to reply to every request can be costly considering communication overheads. A form of silent communication can be adopted in our trading environment, enabling Agents to either refuse to reply to a request or just indicate that they cannot satisfy that request.

Currently, we are working on a formal framework for the specification and logical animation of heterogeneous computational societies. In this framework we define society rules, social roles, social relationships, communication language semantics and social structure. Given such an account of the agent society we will be able to provide a more formal and richer account of the trust modeling of each agent. A multi-agent test-bed [1] is currently being implemented to accommodate our experiments in this formal setting. This test-bed enables the simulation of heterogeneous and possibly antagonistic multi-agent societies and provides a representation of the formally defined social environment of the simulated (trading) communities.

Acknowledgement

This work was partially sponsored by the EC-funded ACTS project AC333 MARINER and partially by the EC-funded ALFEBIITE project (IST-1999-10298).

References

[1] Artikis A., Kamara L. and Pitt J. (2001) "Towards an Open Society Model and Animation", in the Proceedings of Agent-Based Simulation II Workshop, Passau, pp. 48-55

[2] Axelrod, R. (1990) "The Evolution of Cooperation", London: Penguin Books

[3] Barber, K.S. and Kim, J. (2000) "Belief Revision Process Based on Trust: Agents Evaluating Reputation of Information Sources", Workshop on "Deception, Fraud and Trust in Agent Societies", Autonomous Agents, 2000, pp. 15-26

[4] Biswas, A., Debnath, S. and Sen, S. (1999) "Believing Others: Pros and Cons", Proc. IJCAI-99 Workshop "Agents Learning About, From and With Other Agents"

[5] Cantwell, J. (1998) "Resolving Conflicting Information", in the Journal of Logic, Language and Information, Vol. 7, No. 2, pp. 191-220

[6] Castelfranchi C., Conte R., Paolucci M (1998) "Normative Reputation and the Costs of Compliance", in Journal of Artificial Societies and Social Simulation, Vol. 1, No. 3, http://jasss.soc.surrey.ac.uk/1/3/3.html

[7] Castelfranchi, C. and Falcone, R. (1998) "Principles of Trust for MAS: Cognitive Anatomy, Social Importance, and Quantification", ICMAS-98, pp. 72-79

[8] Dignum, F. and Linder, B. (1996) "Modelling Social Agents: Communication as Action", in "Intelligent Agents III: Agent Theories, Architectures, and Languages", pp. 205-217

[9] FIPA (1997) "FIPA-97 Specification, Part 2: Agent Communication Language", http://www.fipa.org

[10] Gambetta, D. (ed.) (1988) "Trust: making and Breaking Cooperative Relations",
      Basil Blackwell, Oxford.
[11] Griffiths, N. and Luck, M. (1999) "Cooperative Plan Selection Through Trust", in
      Proc. 9th European Workshop on Multi-Agent Systems Engineering (MAAMAW
      '99), Springer Verlag, Berlin, pp. 162-174
[12] He, Q, Sycara, K. and Su, Z. (1998) "Security Infrastructure for Software Agent
      Society", in Trust and Deception in Virtual Societies, eds. Castelfranchi and Hua-
      Tan, pp. 137-154
[13] Jennings, N., *et al.* (1999) "FIPA-compliant Agents for Real-time Control of In-
      telligent Network Traffic", Computer Networks, Vol. 31, pp. 2017-2036
[14] Jones, A. and Firozabadi, B. (1998) "On the Characterisation of a Trusting Agent-
      Aspects of a Formal Approach", in Castelfranchi, C. and Hua-Tan (eds.) "Trust
      and Deception in Virtual Societies", pp. 155-166
[15] Jonker, C.M. and Treur, J. (1999) "Formal Analysis of Models for the Dynamics
      of Trust Based on Experiences", in Proc. 9th European Workshop on Multi-Agent
      Systems Engineering (MAAMAW '99), Springer Verlag, Berlin, pp. 221 - 232
[16] Lorenz, E.H. (1988) "Neither Friends nor Strangers: Informal Networks of Sub-
      contracting in French Industry", in: [10], pp. 194-209
[17] Marsh, S. (1994) "Trust in Distributed Artificial Intelligence", in Proc. 4th Euro-
      pean Workshop on Multi-Agent Systems Engineering (MAAMAW '92), Springer
      Verlag, Berlin, pp. 94-113
[18] Patel, A., Prouskas, K., Barria, J. and Pitt, J. (2000) "IN Load Control using a
      Competitive Market-based Multi-Agent System", in Proc. Intelligence and Ser-
      vices in Networks 2000 (IS&N-2000), pp. 239-254
[19] Pham, X.H. and Betts, R. (1994) "Congestion Control in Intelligent Networks",
      Computer Networks and ISDN Systems, Vol. 26, No. 5, pp. 511-524
[20] Prouskas, K., Patel, A., Pitt, J. and Barria, J. (2000) "A Multi-Agent System for
      Intelligent Network Load Control Using a Market-based Approach", in Proc. 4th
      Int. Conf. on MultiAgent Systems (ICMAS-2000), pp. 231-238
[21] Schaerf, A., Shoham, Y. and Tennenholtz, M. (1995) "Adaptive Load Balancing:
      A Study in Multi-Agent Learning", Journal of Artificial Intelligence Research,
      Vol. 2, pp. 475-500
[22] Schillo, M. and Funk, P. (1999) "Learning from and about other Agents in Terms
      of Social Metaphors", in Proc. "Agents Learning about, from and with other
      Agents" Workshop.
[23] Sholtz, F. and Hanrahan, H. (1999) "Market Based Control of SCP Congestion
      in Intelligent Networks", South African Telecommunications Networks and Ap-
      plications Conference (SATNAC-99)
[24] Thirunavukkarasu, C., Finin, T. and Mayfield, J. (1995) "Secret Agents – A
      Security Architecture for the KQML", Proc. ACM CIKM Intelligent Information
      Agents Workshop, Baltimore, December 1995
[25] Watkins, C.J.C.H. (1989) "Learning from Delayed Rewards", King's College,
      Cambridge University (Ph.D. thesis)
[26] Williams, B. (1988) "Formal Structures and Social Reality", in: [10], pp. 3-13
[27] Wong, H.C. and Sycara, K. (1999) "Adding Security and Trust to Multi-Agent
      Systems", in: Proc. of Autonomous Agents '99 Workshop on Deception, Fraud,
      and Trust in Agent Societies, pp. 149-161
[28] Zacharia, G. (1999) "Trust Management Through Reputation Mechanisms",
      Workshop on "Deception, Fraud and Trust in Agent Societies", Autonomous
      Agents, 1999, pp. 163-167

Learning to Trust

Andreas Birk

Vrije Universiteit Brussel, Artificial Intelligence Laboratory,
Pleinlaan 2, 10G725, 1050 Brussels, Belgium
birk@ieee.org

Abstract. Evolutionary game-theory is a powerful tool to investigate the development of complex relations between individuals such as the emergence of cooperation and trust. But the propagation of genes is an unrealistic assumption when it comes to model fast-changing social interactions. We show how a transition from evolutionary game-theory to learning can be made. Specifically, we show how cooperation and trust can develop together through social interactions and a suited learning mechanism.

1 Introduction

Evolutionary game theory [Axe84,Smi82] is a powerful tool for the investigation of interactions between individuals. It has especially become popular with research on cooperation (see e.g. [AD94] for an overview), but it also has been applied to many other domains.

Evolutionary methods are built upon a transfer of encoded information, i.e., genes, between individuals. This transfer of genes includes two main assumptions. The first one is the "feasibility of breeding" assumption. Evolution includes the generation of new individuals and the deletion or death of others. The second one is the "obey mother nature assumption". When an off-spring is generated, it has no choice whether to incorporate a particular gene or not; the "decision" is made by mother nature or by stochastic operators in simulated evolution.

Let us examine these assumptions from the perspective of trust as a social phenomenon. On the one hand, genetic evolution is very likely to influence animal and especially human behavior also in respect to social interactions. On the other hand, social developments happen on a completely different time-scale than evolution. Therefore, it should be clear that evolution can not serve as the only explanation. The work presented here shows a possible way out of the problems with evolutionary schemes. Instead of evolution, a learning approach somewhat in the spirit of selectionism [Ede87,Ede85] is used here. The main feature of this learning algorithm is that it is based on a pool of potential solutions or so-called hypotheses in each individual. This type of algorithm can also be highly efficient for the learning of individual skills as demonstrated in experiments with learning eye-hand coordination in simulations and real robot systems [BP00,Bir96] and experiments on learning several basic behaviors in a robotic ecosystem [Bir98].

R. Falcone, M. Singh, and Y.-H. Tan (Eds.): Trust in Cyber-societies, LNAI 2246, pp. 133–144, 2001.
© Springer-Verlag Berlin Heidelberg 2001

This paper builds on previous results based on evolutionary game theory [Bir00]. It is shown here that cooperation and trust in a continuous-case N-player prisoner's dilemma can not only evolve, but can also be learned. In the general framework of this research, trust is based upon dynamical processes. Other commonly used notions of trust build upon compliance-based approaches, using for example standardized protocols and cryptography. The dynamical notion of trust which is used here guarantess no "absolute" security as trusted systems can cheat. But the process has the important advantage of being completely open and robust. A selection of different approches to trust can be found for example in [CCe00].

In this research framework, the basis for trust is seen as an intrinsic property of each individual in form of the so-called *trustworthiness*. The trustworthiness of an individual a_A is an objective measure for another individual a_B of the desirability of interactions with a_A. If the, possibly continuous, trustworthiness of a_A is high, it is highly desirable for a_B to engage in trust-based interactions with a_A. The trustworthiness of a_A can be dynamic and it is not directly perceivable by a_B. Any process which tries to establish an approximation of the trustworthiness of a_A is denoted as building *trust* in this research framework.

Processes for building trust often include a non-rational component in the sense that decisions on how to deal with another individual are not only based on previous interactions with this individual, but also on other subjective criteria, such as outer appearance, recommendations from others, and so on. Subjective processes for building trust are extremely important as they allow decisions whether to interact or not with unknown individuals, i.e., individuals who have not been encountered in previous interactions. Labels, which do not bear meanings in the beginnings of the experiments, and preferences to interact with individuals carrying certain labels are used here to model the building of trust.

The rest of this paper is structured as follows. In section 2, the basic ideas of the transition from evolutionary game theory to social learning are explained. A continuous case N-players prisoner's dilemma is introduced in section 3 as a basis for the experimental framework. In section 4, the concrete learning algorithms and results are presented. Section 5 concludes the paper.

2   Selectionist Learning within Individual Minds

Evolutionary algorithms with their major classes Genetic Algorithms [Gol89], [Hol75], Evolutionary Programming [FOW66], Evolutionary Strategies [Sch77], [Rec73] and Genetic Programming [Koz94b,Koz92], imitate, or at least are inspired by, the principle of evolution in nature. They use a set of potential solutions (the *population*) to a particular problem or *domain*. Populations are generated in iterations (*generations*) using operations for *selection* and *transformation*. In doing so, the selection and transformation operations focus on good, in respect to a *fitness* function, members of the population. As better members are more likely to be chosen an improvement over time is expected.

For the sake of simplicity, we refer here to any representation of a potential solution as a *gene*, typically a fixed-length binary string or a parse-tree. When using evolutionary algorithms to investigate artificial "living" systems as in the fields of evolutionary game theory [Axe84,Smi82] or evolutionary robotics [GHF94,WFP99,DBB98,FM94,Koz94a], a single gene determines a crucial aspect of an individual system, such as its morphology [Sim94], its control [FM94], or highlevel behavior like a strategy in social interactions [AH81].

Here we propose a mechanism which is not based on evolution, but which is a learning mechanism inspired by the evolutionary driving forces of selection and the generation of diversity, somewhat in the spirit of selectionism [Ede87,Ede85]. Here a crucial aspect of an individual system is not determined by a fixed gene but by a so-called *hypothesis*. For example, in the domain of robot-control, a certain hypothesis h would for example represent that given a situation s the behavior b would be appropriate. The crucial aspect of a hypothesis is that, unlike in the case of a gene, there is not a single hypothesis for a particular problem. Instead, an individual has *multiple hypotheses* about potential solution for a single instance from the domain. In the case of robot-control, this means that given a situation s there is a set of hypotheses HS linking s to several possible behaviors.

Hypotheses are ranked within the individual by a so-called *preference* function $pref()$. The best hypothesis according to this ranking is most likely to be activated, e.g., to be expressed as an behavior or to serve as a (partial) model of the world. Lower ranking hypotheses also have a chance to become activated. The retrieval of the hypothesis which becomes active can for example be done with the roulette-wheel (RW) principle as follows. Given a hypotheses-set HS and the preferences $pref(h)$ for all h in HS, the likelihood $prob$ that a particular hypothesis h' is retrieved from HS for activation is proportional to its preference, i.e.,

$$prob(h' \text{ is activated}) = pref(h') / \sum_{h \in HS} pref(h)$$

Note that it is important not to confuse RW-retrieval with RW-selection from evolutionary algorithms. In the case of evolutionary selection, a gene is transferred into the next generation. If this does not happen, the gene dies out, i.e., it disappears from the population. When a hypothesis h is selected by RW-retrieval, it is applied and tested. This does not necessarily result in a change in the set HS of hypotheses with which h is in concurrency.

Through the retrieval and activation of h, information about the usefulness of h is gathered and $pref(h)$ is updated. This in turn can lead to changes of h and even its elimination, but as mentioned above, this is not necessarily the case. With evolution, potential solutions are encoded in genes and transferred among individuals as generations progress. With multiple-hypotheses learning, the potential solutions in form of hypotheses are never transfered between different individuals. Nevertheless, similar hypotheses-sets in different individuals and coordinated usage of hypotheses can emerge through suited social interactions as will be shown later on in experiments.

As already mentioned in the introduction, variations of the learning algorithm presented here have also been applied to learning of individual skills on the level of sensor-motor control and on the level of behaviors for robots in real world environments [BP00,Bir96,Bir98]. In these experiments, it has been shown that a pool of potential solutions in the "mind" of a single individual increases the robustness against distortions from real world noise and it can increase the learning speed through the re-use of partial solutions found in the pool.

3   The Experimental Framework

3.1   A Continuous-Case N-Player Prisoner's Dilemma

The basis for the experiments described later on is a version of the prisoner's dilemma with N players and continuous cases of investment and payoffs (CN-PD). It is motivated and described in more detail in [Bir00,BW00].

Each agent a_i has a so-called cooperation-level $co_i \in [0.0; 1.0]$. In a game, the cooperation-level determines the agent's investment I_i, which serves for the benefit of the group (including a_i itself). Concretely, the investment is determined by:

$$I_i = co_i \cdot 75$$

Let $\bar{co}$ denote the average cooperation-level of the group, i.e.:

$$\bar{co} = \sum_{1 \le i \le N} co_i / N$$

The so-called gain G_i for an agent a_i is determined by:

$$G_i = \bar{co} \cdot 100$$

Roughly, all investments are collected, some profit is generated with the investments, and finally investments and profit are distributed among the investors. The dilemma arises as investments and profit are equally shared among all. Thus there is the temptation to invest less than the others and to exploit their contribution to the profit. This becomes even clearer when we look at the netgain or payoff for each agent. This payoff po_i for an agent a_i is the difference between gain and investment, i.e.:

$$po_i = G_i - I_i = \sum_{1 \le j \le N} co_j / N \cdot 100 - co_i \cdot 75$$

So, on the one hand, it is in the interest of each agent that there is a high overall investment. On the other hand, there is the temptation to leave the task of investing to others, as the overall gain is distributed among all, independent of the individual investment. Note, that the payoff for an agent depends on its own cooperation level co_i and on the average cooperation level $\bar{co}$. Its profit function $f_p : [0,1] \times [0,1] \to I\!R$ is thus

$$f_p(co_i, \bar{co}) = co_i \cdot -75 + \bar{co} \cdot 100$$

Based on this, we can extend the terminology for payoff values in the standard prisoner's dilemma, with payoff types for cooperation (C), punishment (P), temptation (T), and sucking (S), as follows:

- Full cooperation as all fully invest: $C_{all} = f_p(1.0, 1.0) = 25$
- All punished as nobody invests: $P_{all} = f_p(0.0, 0.0) = 0$
- Maximum temptation: $T_{max} = f_p(0.0, \frac{N-1}{N}) \geq 50$
- Maximum sucking: $S_{max} = f_p(0.0, \frac{1}{N}) \leq -25$

For $co, \bar{co} \neq 0.0, 1.0$, we get the following additional types of payoffs, the so-called partial temptation, the weak cooperation, the single punishment, and the partial sucking. They are not constants (for a fixed N) like the previous ones, but actual functions in $(co, \bar{co})$. Concretely, they are sub-functions of $f_p(co, \bar{co})$, operating on sub-spaces defined by relations of co in respect to $\bar{co}$.

3.2 Strategies for Iterated Games

When playing iterated games, the concept of a strategy [Axe84] can be used to determine the behavior of an agent. This means, the outcome of previous games is used to compute whether to cooperate or not in the recent game, or to compute the degree of cooperation in the continuous case [RS98].

In [BW00] it is shown that the so-called justified snobism (JS) is a successful strategy for the continuous case N-player prisoner's dilemma. JS cooperates slightly more than the average cooperation level of the group of N players if a non-negative payoff was achieved in the previous iteration, and it cooperates exactly at the previous average cooperation level of the group otherwise.

Justified-Snobism (JS):
$$po_i(t-1) \geq 0 : co_i(t) = \bar{co}(t-1) + c_{JS}$$
$$po_i(t-1) < 0 : co_i(t) = \bar{co}(t-1)$$

So, JS tries to be slightly more cooperative than the average. This leads to the name for this strategy as the snobbish belief to be "better" (in terms of altruism) than the average of the group is somehow justified for players which use this strategy.

In addition, following strategies are used in the experiments described lateron to challenge JS:

Follow-the-masses (FTM) : match the average cooperation level from the previous iteration, i.e., $co_i[t] = \bar{co}[t-1]$

Hide-in-the-masses (HIM) : subtract a small constant c from the average cooperation level, i.e., $co_i[t] = \bar{co}[t-1] - c$

Occasional-short-changed-JS (OSC-JS) : a slight variation of JS, where occasionally a small constant c is subtracted from the JS investment

Occasional-cheating-JS (OC-JS) : an other slight variation of JS, where occasionally nothing is invested

Challenge-the-masses (CTM) : Zero cooperation when the previous average cooperation is below one's cooperation level, a constant cooperation level c' otherwise, i.e.,
- $co_i[t-1] \geq \bar{co} : co_i[t] = c'$
- $co_i[t-1] < \bar{co} : co_i[t] = 0$

Non-altruism (NA) : always completely defect, i.e., $co_i[t] = 0$

Anything-will-do (AWD) : always cooperate at a fixed level, i.e., $co_i[t] = c'$

In evolutionary game theory, each agent has exactly one strategy, which is encoded in a single gene. The survival of the agent and the number of its off springs depend on the performance of this strategy. Here, each agent has a set of strategies from which he can choose. This means strategies are encoded as multiple-hypotheses. The set of strategy-hypotheses for an agent a_i is denoted with HS_i^s.

When playing games, an agent must first retrieve a strategy s from HS_i^s. This is done using roulette-wheel retrieval as introduced above. The outcome of the game is then used to update the preference $pref(h)$ for the hypothesis that strategy s is useful for getting a high payoff in a game.

3.3 The Basis of Trust

Much like in [Bir00], trust here is expressed by preferences of an agent to be grouped together with certain other agents in a game. Again, subjective criteria in the form of labels, as a kind of outer appearance of agents, is used for this. The two major differences with previous work are that here

- the emergence of cooperation and trust is grounded in learning instead of using evolution, and
- agents can change their labels, i.e., there is no a priori assignments of labels to individuals.

The basic principle of the so-called trust-function stays the same as in previous work. So, the function $trust_i : L \to [0.0, 1.0]$ of an agent a_i maps a weight w to each possible label l_j, such that $trust_i(l_j) = w$. The weight w represents a_i's preference to interact with an agent with label l_j. If w is high, i.e., close to 1.0, a_i prefers to interact with agents with label l_j, or it simply trusts them. If w is low, i.e., close to 0.0, a_i prefers not to interact with agents with label l_j, or it simply does not trust them.

As mentioned above, individuals do not have a fixed label in the experiments reported here. Instead, the labels are represented within each agent as multiple-hypotheses. The set of label-hypotheses for an agent a_i is denoted with HS_i^l. Before each game, an agent must decide which label it signals to the other agents. This will influence the formation of groups and the outcome of the games. So choosing a label is a hypothesis-retrieval and hypothesis-activation for an agent. The hypothesis is that when a_i signals this particular label, the outcome of the next game will be beneficial for a_i.

```
1   form group G {
2   /* randomly initialize the group G with one agent */
3       G = ∅
4       S_temp = S
5       a = random select (S_temp)
6       G = G ∪ {a}
7       S_temp = S_temp/{a}
8   /* add agents to G based on the trust of the agents already in G */
9       while #G < N
10          ∀l_i ∈ S_L : sw(l_i) = ∑_{a_j ∈ G} trust_j(l_i)
11          a' = roulette-wheel selection (S_temp, sw())
12          G = G ∪ {a'}
13          S_temp = S_temp/{a}
14      }
15  }
```

Fig. 1. Group formation based on the trust-functions.

4 Learning Strategies, Signals, and Trust

4.1 The Structure of the Iterated Games

Before the algorithm running within an agent is presented in more detail, let us first have another look at the overall game. There is a set of agents with a fixed cardinality n_S, the so-called *society* S. The society plays iterated CN-PDs in time-steps t. At the beginning of each time-step, the society is split into groups of size N biased by the individual trust-functions.

The concrete algorithm for doing this is shown in Figure 1. First, the group is randomly initialized with one agent. Then, additional agents are put into the group. In doing so, the likelihood of placing an agent a_i who signals label l into the group is proportional to the summed trust in l of all agents which are already in the group. Note that agents are true individuals in the sense that each agent can only be once in one particular group during a game. In evolutionary games in contrast, agents can multiply by generating offspring and thus be represented several times in several groups.

After the groups are formed, several CN-PD games are played and payoffs for each agent are generated. The payoffs are used to update the preferences of the different hypotheses as will be shown in the next section. Afterwards, the groups are mixed together into a uniform society again and the overall process proceeds to the next time-step, $t + 1$.

```
1   behavior agent aᵢ in game G[t] {
2       RW-retrieve label l
3       signal l      /* and the agent is placed in a group */
4       RW-retrieve strategy s
5       play s
6   /* update preferences for labels */
7       if poᵢ ≥ 0
8           pref(l)[t] = q · pref(l)[t − 1](1 − q) · poᵢ
9       else
10          ∀l′ ∈ HSⁱᵢ/{l} : pref(l′)[t] = q · pref(l′)[t − 1](1 − q) · |poᵢ|
11  /* update preferences for strategies */
12      if poᵢ ≥ 0
13          pref(s)[t] = q · pref(s)[t − 1](1 − q) · poᵢ
14      else
15          ∀s′ ∈ HSˢᵢ/{s} : pref(s′)[t] = q · pref(s′)[t − 1](1 − q) · |poᵢ|
16  /* update trust-function */
17      trustᵢ(lⱼ)[t] = (1 − q) · trustᵢ(lⱼ)[t − 1]
18          +q · poᵢ[t − 1] · #{aₖ ∈ G with L(aₖ) = lⱼ}/N_A
19  }
```

Fig. 2. The behavior of an individual agent in a single game in a Pseudo-Code.

4.2   The Algorithm within an Agent

Figure 2 shows the algorithm running within an individual agent. Most of it has been explained and motivated above. What remains to be defined is the update of the preferences of different hypotheses (lines 7 to 10 for labels and lines 12 to 15 for strategies).

The main idea for the update is simply that the running average of payoffs is used as indication of how preferrable a certain strategy or label is. As a minor problem, negative payoffs have to be taken care of. To ensure that the preferences never become negative as the standard roulette-wheel principle is only applicable with positive weights. Therefore, negative payoffs do not decrease the preference for a strategy s, which was active in the last time-step, but they lead to an increase in the preference of all other strategies except s (line 10). The same holds in respect to labels (line 15).

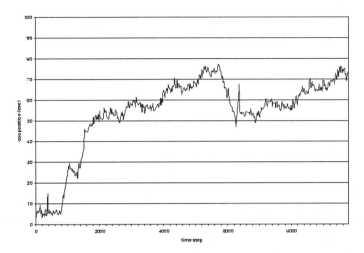

Fig. 3. The simultaneous learning of suited strategies, signaled labels, and trust-functions leads to an emergent cooperation.

4.3   Results

In the experiments reported here, the size n_S of the society is 100, the group sizes are always 10 agents. After grouping, the agents always play 50 games together to collect payoffs before they proceed to the next time-step and new groups are formed. The preferences for the different hypotheses are randomly initialized for both types, i.e., for strategies as well as for labels.

As shown in figure 3, the simultaneous learning of strategies, signaled labels, and trust-functions leads to an emergent cooperation, i.e., an increasing average cooperation level. This cooperation is nevertheless very vulnerable. Unlike evolutionary experiments where the JS as strategy and the trust into a particular label become dominant and stable, the multiple-hypotheses learning leads to dynamic scenarios.

Especially, when the preferences for JS and certain labels become very high and stable, some agents will (re-)discover in their hypotheses-sets HS^s and HS^l non-altruistic strategies and how to use deceptive labels. Note, that unlike in the genetic case, these "bad behaviors" can never die out. They are always present, even if they are sometimes rarely observered due to small preferences.

So, when the preferences for JS and certain labels become very high and stable in the majority of the society, activating non-altruistic behaviors and deceptive signals is very profitable as a high amount of general goodwill can be exploited. Agents which retrieve these hypotheses by chance receive high payoffs and therefore increase their preference for them substantially. Therefore, they are likely to retrieve and activate them in the next round again, and so on.

A substantial break in the general cooperativeness and trust can hence occasionally be observed. In a snowball reaction, more and more agents discover "bad behaviors". But the more agents do so, the less profitable this becomes. Therefore, the society is then able to recover again.

5   Conclusion

The work presented in this paper is set in a research context where trust is modeled as dynamic preferences of whether to engage in social interactions with others. The basis of this model is an intrinsic property called trustworthiness in every individual a. Trustworthiness of a is an objective measure for other individuals regarding whether it is desirable to engage in an interaction with a. But trustworthiness cannot directly be perceived. Building trust therefore relates in this model to the estimation of trustworthiness. Subjective criteria like the outer appearance are important for building trust as they allow to handle unknown agents for whom data from previous interactions does not exist. Hence, trust is represented as preference to be grouped together with agents with a certain label to play a game.

In previous work, it was shown that stable relations of trust can evolve and that the co-evolution of trust can boost the evolution of cooperation. In general, evolutionary game-theory is a well known tool for investigating basic properties of interactions between individuals. But the transfer of encoded information, i.e., genes, is unsuited as main basis for models of social interactions. Social interactions do not operate on the time-scale of natural evolution nor do they provide such powerful means of information exchange as the transfer of genes. Here, we show how learning can be used to overcome this severe drawback. The so-called multiple-hypotheses approach is used to successfully develop cooperation and trust simultaneously in scenarios modeled by a continuous-case N-player prisoner's dilemma.

Acknowledgments

Andreas Birk is a research fellow (OZM-980252) of the Flemish Institution for Applied Research (IWT).

References

[AD94]   Robert Axelrod and Lisa D'Ambrosio. An annotated bibliography on the evolution of cooperation. *http://www.ipps.lsa.umich.edu/ipps/papers/coop/ Evol_of_Coop_Bibliography.txt*, October 1994.

[AH81]   R. Axelrod and W. D. Hamilton. The evolution of cooperation. *Science*, 211:1390–1396, 1981.

[Axe84]   R. Axelrod. *The Evolution of Cooperation*. Basic Books, 1984.

[Bir96]   Andreas Birk. Learning geometric concepts with an evolutionary algorithm. In *Proc. of The Fifth Annual Conference on Evolutionary Programming*. The MIT Press, Cambridge, 1996.

[Bir98]   Andreas Birk. Robot learning and self-sufficiency: What the energy-level can tell us about a robot's performance. In *Proceedings of the Sixth European Workshop on Learning Robots*, LNAI 1545. Springer, 1998.

[Bir00]   Andreas Birk. Boosting cooperation by evolving trust. *Applied Artificial Intelligence Journal*, 14(8), September 2000.

[BP00]   Andreas Birk and Wolfgang J. Paul. Schemas and genetic programming. In Ritter, Cruse, and Dean, editors, *Prerational Intelligence*, volume 2. Kluwer, 2000.

[BW00]   Andreas Birk and Julie Wiernik. A successful strategy for a real-world instance of the n-player prisoner's dilemma with continuous degrees of cooperation. Technical report, Vrije Universiteit Brussel, AI-Laboratory, 2000.

[CCe00]   Babak Sadighi Firozabadi Cristiano Castelfranchi, Rino Falcone and Yao Hua Tan (eds). Special issue: Trust in agents. *Applied Artificial Intelligence Journal*, 14(8), September 2000.

[DBB98]   Peter Dittrich, Andreas Burgel, and Wolfgang Banzhaf. Ra *Robots and Autonomous Systems*, 1998.

[Ede85]   Gerald M. Edelman. Neural darwinism: Population thinking and higher brain function. In Michael Shafto, editor, *How We Know*, pages 1–30. Harper and Row, 1985.

[Ede87]   Gerald M. Edelman. *Neural Darwinism: The Theory of Neuronal Group Selection*. Basic Books, New York, 1987.

[FM94]   D. Floreano and F. Mondada. Automatic creation of an autonomous agent: Genetic evolution of a neural-network driven robot. In *Proceedings of the Conference on Simulation of Adaptive Behavior*, 1994.

[FOW66]   L.J. Fogel, A.J. Owens, and M.J. Walsh. *Artificial Intelligence through Simulated Evolution*. Wiley, New York, 1966.

[GHF94]   R. Ghanea-Hercock and A. P Fraser. Evolution of autonomous robot control architectures. In T. C. Fogarty, editor, *Evolutionary Computing*, Lecture Notes in Computer Science. Springer-Verlag, 1994.

[Gol89]   David Goldberg. *Genetic Algorithms in Search Optimization and Machine Learning*. Addison-Wesley, Reading, 1989.

[Hol75]   John H. Holland. *Adaptation in Natural and Artificial Systems*. The University of Michigan Press, Ann Arbor, 1975.

[Koz92]   John R. Koza. *Genetic programming*. The MIT Press, Cambridge, 1992.

[Koz94a]   John R. Koza. *Evolution of a subsumption architecture that performs a wall following task for an autonomous mobile robot*, volume II: Intersections Between Theory and Experiment, chapter 19, pages 321–346. MIT Press, 1994.

[Koz94b]   John R. Koza. *Genetic programming II*. The MIT Press, Cambridge, 1994.

[Rec73]   Ingo Rechenberg. *Evolutionsstrategie: Optimierung technischer Systeme nach Prinzipien der biologischen Evolution*. Fromman-Holzboog, Stuttgart, 1973.

[RS98]   Gilbert Roberts and Thomas N. Sherratt. Development of cooperative relationships through increasing investment. *Nature*, 394 (July):175–179, 1998.

[Sch77]   Hans Paul Schwefel. *Numerische Optimierung von Computer-Modellen mittels der Evolutions-Strategie*. Birkhäuser, Basel, 1977.

[Sim94]   Karl Sims. Evolving 3D morphologgy and behavior by competition. In R. Brooks and P. Maes, editors, *Artificial Life IV*, pages 28–39, Cambridge, MA, 1994. MIT Press.

[Smi82]   J. Maynard Smith. *Evolution and the Theory of Games*. Cambridge University Press, Cambridge, 1982.

[WFP99] Richard A. Watson, Sevan G. Ficici, and Jordan B. Pollack. Embodied evolution: Embodying an evolutionary algorithm in a population of robots. In Peter J. Angeline, Zbyszek Michalewicz, Marc Schoenauer, Xin Yao, and Ali Zalzala, editors, *Proceedings of the Congress on Evolutionary Computation*, volume 1, pages 335–342. IEEE Press, 1999.

Learning Mutual Trust

Rajatish Mukherjee, Bikramjit Banerjee, and Sandip Sen

Mathematical & Computer Sciences Department, University of Tulsa
{rajatish,bikram,sandip}@euler.mcs.utulsa.edu
http://www.mcs.utulsa.edu/~sandip

Abstract. The multiagent learning literature has looked at iterated two-player games to develop mechanisms that allow agents to learn to converge on Nash Equilibrium strategy profiles. An equilibrium configuration implies that there is no motivation for one player to change its strategy if the other does not. Often, in general sum games, a higher payoff can be obtained by both players if one chooses not to respond optimally to the other player. By developing mutual trust, agents can avoid iterated best responses that will lead to a lesser payoff Nash Equilibrium. In this paper we consider 1-level agents (modelers) who select actions based on expected utility considering probability distributions over the actions of the opponent(s). We show that in certain situations, such stochastically-greedy agents can perform better (by developing mutually trusting behavior) than those that explicitly attempt to converge to Nash Equilibrium. We also experiment with an interesting action revealation strategy that can give the revealer better payoff on convergence than a non-revealing approach. By revealing, the revealer enables the opponent to agree to a more trusted equilibrium.

1 Introduction

Reinforcement learning techniques with performance and convergence guarantees have been developed for isolated single agents [13]. The underlying assumption of such a proof is that the environment is stationary. Multiagent or concurrent learning, however, violates this assumption. As a result, the standard reinforcement learning techniques (like Q-learning) are not guaranteed to converge in a multiagent environment. The desired convergence in multiagent systems is to an equilibrium strategy-profile (collection of strategies of the agents) rather than optimal strategies for an individual agent.

The stochastic-game (or *Markov Games*) framework, a generalization of Markov decision processes for multiple players, has been used to model learning by agents in various domains [2, 5, 4]. In [2], two basic types of multiagent learners have been studied. The learners who do not model other agents, effectively considering them as parts of a non-stationary environment, are called 'independent learners' (ILs). We term these 0-level agents. In contrast to such agents, those that observe others' actions and rewards and use these explicitly in modeling them, are called 'joint-action learners' (JALs). We call these 1-level

R. Falcone, M. Singh, and Y.-H. Tan (Eds.): Trust in Cyber-societies, LNAI 2246, pp. 145–157, 2001.
© Springer-Verlag Berlin Heidelberg 2001

agents. Theorem 1 in [2] claims that both 0 and 1-level agents converge to equilibria in coordination games. But their work is not extendible to general domains (general-sum games). Hu and Wellman have adopted a complete-information general-sum game approach and provide a learning scheme that allows learners to converge to a mixed-strategy Nash Equilibrium in the limit [4].

Nash Equilibrium, however, does not guarantee that agents will obtain the best possible payoffs. Our aim is to determine if the agents can learn to converge to a more desirable pareto-optimal solution than a Nash Equilibrium. Some non-Nash Equilibrium action combinations may yield better payoffs for both agents, which may be reached if the agents look ahead while selecting actions [1]. Such desirable non-myopic choices are preferred by both agents. While playing best response to other agents' current policy will lead to a deviation from such desirable solutions, restraint or mutual trust can enable players to stick to such action combinations.

In this paper we evaluate the possibility of concurrent learners converging to such desirable non-myopic action choices. While Hu and Wellman's approach is guaranteed to converge to Nash Equilibrium strategy profiles [4], independent, or even ordinary 1-level Q-learners have no such guarantees. In previous work, we observed that 0-level Q-learners often outperformed higher-level Q-learners in the long run even though their learning rate is slower [8]. In this paper we show that greedy modelers can, in their turn, outperform equilibrium seeking modelers in terms of the rewards received.

2   Definitions

In this section, we introduce some definitions to formulate a framework for concurrent learning.

Definition 1 *A Markov Decision Process (MDP) is a quadruple $\{S, A, T, R\}$, where S is the set of states, A is the set of actions, T is the transition function, $T : S \times A \rightarrow PD(S)$, PD being a probability distribution, and R is the reward function, $R : S \times A \rightarrow \mathcal{R}$.*

A multiagent reinforcement-learning task can be viewed as an extended MDP, with S specifying the joint-state of the agents, A being the joint-actions of the agents, $(A_1 \times A_2 \times \ldots A_n$ where A_i is the set of actions avaiable to the ith agent), T as the joint state-transition function, and the reward function is redefined as $R : S \times A \rightarrow \mathcal{R}^n$. The functions T and R are usually unknown, necessitating learning. A strategy is a probability matrix over the action space. The goal of the ith agent is to find a strategy π_i that maximizes its expected sum of discounted rewards,

$$v(s, \pi_i) = \sum_{t=0}^{\infty} \gamma^t E(r_t^i | \pi_i, \pi_{-i}, s_0 = s)$$

where s_0 is the initial joint-state, r_t^i is the reward of the ith agent at time t, $\gamma \in [0, 1)$ is the discount factor, and π_{-i} is the strategy-profile of i's opponents.

In [4] the *ith* agent learns π_{-i} simultaneously, and opts for the best response to it. Though myopically this is the best an agent can do, it may miss opportunities for receiving higher payoffs as in the well-known Prisoner's Dilemma problem [11].

Definition 2 *A bimatrix game is given by a pair of matrices, (M_1, M_2), (each of size $|A_1| \times |A_2|$) for a two-agent game, where the payoff of the ith agent for the joint action (a_1, a_2) is given by the entry $M_i(a_1, a_2)$, $\forall(a_1, a_2) \in A_1 \times A_2$, $i = 1, 2$.*

Each stage of an extended-MDP for two agents (it can be extended to n agents using n-dimensional tables instead of matrices), can be looked upon as a bimatrix game. A *zero-sum game* is a special bimatrix game where $M_1(a_1, a_2) + M_2(a_1, a_2) = 0$, $\forall(a_1, a_2) \in A_1 \times A_2$. In this paper we consider general-sum games, where the above sum is not a constant, and hence the individual payoffs of the agents for any joint-action are uncorrelated. We now define Nash equilibrium for such games.

Definition 3 *A pure-strategy Nash Equilibrium for a bimatrix game (M_1, M_2) is a pair of actions (a_1^*, a_2^*) such that*

$$M_1(a_1^*, a_2^*) \geq M_1(a_1, a_2^*) \quad \forall a_1 \in A_1$$

$$M_2(a_1^*, a_2^*) \geq M_2(a_1^*, a_2) \quad \forall a_2 \in A_2$$

In a Nash equlibrium the action chosen by each player is the best response to the opponent's current strategy and no player in this game has any incentive for unilateral deviation from its current strategy. A general-sum bimatrix game may not have any pure-strategy Nash Equilibrium.

Definition 4 *A mixed-strategy Nash Equilibrium for a bimatrix game (M_1, M_2) is a pair of probability vectors (π_1^*, π_2^*) such that*

$$\pi_1^* M_1 \pi_2^* \geq \pi_1 M_1 \pi_2^* \quad \forall \pi_1 \in PD(A_1)$$

$$\pi_1^* M_2 \pi_2^* \geq \pi_1^* M_2 \pi_2 \quad \forall \pi_2 \in PD(A_2)$$

where $PD(A_i)$ is the set of probability-distributions over the action space of the ith agent.

A significant property of mixed-strategy Nash Equilibria, is that there always exists at least one such equilibrium profile for an arbitrary finite bimatrix game [9]. Given such a bimatrix game (M_1, M_2), the mixed-strategy Nash Equilibrium, (π_1^*, π_2^*), can be computed using a quadratic programming approach as outlined in [7].

Researchers have worked with other kinds of noncooperative equilibrium, e.g., correlated equilibrium [3], . We are interested in a non-myopic equilibrium where a player not only considers its best response to current playing trends, but also future possible retaliations by the other player. For example, consider the two players playing π_1^A and π_1^B respectively and the first player getting $\pi_1^A M_1 \pi_1^B$ as a result. While considering another strategy π_2^A, A now considers not only if $\pi_2^A M_1 \pi_1^B > \pi_1^A M_1 \pi_1^B$, but also if $\pi_2^A M_1 \pi_1^B > \pi_2^A M_1 \pi_2^B$, where π_2^B is Bs best response to π_2^A (this equilibrium concept is similar in motivation to the non-myopic equilibrium in the Theory of Moves approach [1]). Of course, it is difficult to estimate the other player's best response, but this can be approximated based on past play of the opponent.

3   Reinforcement-Learning

A general, single-agent reinforcement learning task is an MDP, where the state transition and reward functions T and R are unknown. A simple, model-free and on-line technique for reinforcement learning is Q-learning [14]. In a stateless domain, as is the case with single-stage games studied in this paper, an independent Q-learner will have Q-values for each action a, $Q(a)$, and update them based on rewards r received from taking action a as follows:

$$Q(a) \leftarrow Q(a) + \alpha(r - Q(a))$$

where α is the learning-rate. This iteration has been proved to converge to optimal Q-values, for a particular structure of α, but independent of any particular exploration strategy provided it satisfies some general requirements. When a number of independent learners apply this algorithm, the convergence-guarantee does not hold due to the non-stationarity of the environment. However, such straightforward applications of Q-learning in multiagent systems have achieved success in the past [2, 10, 12, 15]. Our 1-level Q-learners learn Q-values, $Q(a, b)$, for each possible joint-action (a, b), using its observation of the actions of the other agents, but solely its own reward for joint-action. Thus the updation-rule used is

$$Q(a, b) \leftarrow Q(a, b) + \alpha(r - Q(a, b))$$

To allow these 1-level Q-learning agents to increasingly exploit their learned strategies, we use the Boltzmann exploration strategy, which slowly increases the exploitation probability. In this exploration scheme, the action a is selected with probability

$$\frac{e^{E(Q(a))/T}}{\sum_{a'} e^{E(Q(a'))/T}}$$

where $E(a) = \sum_b p_b Q(a, b)$, p_b being computed as the relative-frequency measure from B's action history. Thus we call these agents "expected utility based probabilistic learners" or (EUPs). The temperature parameter T is started at a high value (causing more exploration) and then decreased over time, e.g., by muliplying with a decay factor, to increase the exploitation probability.

We have also experimented with an interesting variation of simulataneous play. We allow one player to reveal or announce its move in each game. The revealing player can either be fixed (unilateral revelation) or randomly chosen from the two players (bilateral revelation). The other player can choose its move based on complete knowledge by the move made by its opponent. It might still decide to explore its actions instead of playing best response in order to thoroughly evaluate its options. In the revealing version of the game, the revealing player(s) keep not only an estimate of p_b, the frequency distribution of its opponent's moves, but also the corresponding conditional frequency distribution, $p_{b|a}$, i.e., the likelihood that the opponent is going to play its move b if the revealer plays b. The revealer reveals its moves, a, with probability

$$\frac{e^{\bar{E}(Q(a))/T}}{\sum_{a'} e^{\bar{E}(Q(a'))/T}}$$

where $\bar{E}(a) = \sum_b p_{b|a} Q(a,b)$. In response, the other player can play b with probability

$$\frac{e^{E(Q(a,b)/T}}{\sum_{b'} e^{E(Q(a,b'))/T}}$$

4   Experiments

Our experimental work uses four game matrices (Figures 1, 3, 5 and 7) to highlight how the agents learn to trust each other. We experiment with 3×3 game matrices. Each agent has three actions to choose from, where a_is are the actions of agent A and b_is those of agent B. For any action combination, the top-right value in the corresponding matrix cell is the payoff to agent B and the bottom-left value is the payoff to agent A. The shaded entry in each matrix corresponds to the Nash Equilibrium strategy-profile. The action-profile that the agents prefer (greedy) and the desirable non-myopic solutions are also marked in each game-matrix.

Depending on the payoffs in the matrices, the agents foresee sufficient rewards to reveal their actions to the other agents. For example, consider matrix in figure 7. This matrix is designed to demonstrate how revelation of an agent's (agent A) actions (while the other agent withholds such information) may lead to less payoff for the agent. However, when both of the agent's reveal their information, they learn to avoid the greedy solution and converge to a more "desirable" solution. Similarly, the matrix in figure 1 is used to show the failure of bilateral revelation of information.

In figure 1 (left) there is a single pure Nash Equilibrium given by the action-profile $\langle a_3, b_3 \rangle$ giving a payoff of 5 to both agents. The desirable solution, however, is for the action-combination $\langle a_1, b_1 \rangle$ giving a payoff of 10 to both agents. We used two EUPs using the above Q-learning algorithm, learning for 1000 iterations and using 0.99 as the temperature decay factor starting at $T = 10$. The probabilities of adopting joint-actions $\langle a_1, b_1 \rangle$ and $\langle a_3, b_3 \rangle$ as measured by frequencies were recorded every 100 interactions averaged over the last 100 interactions. The values in the figures were averaged over 10 runs, and these probabilities are plotted in figure 1 (right). In this case, the EUPs converge to the Nash Equilibrium in most of the runs even though the payoff is less than the

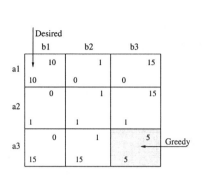
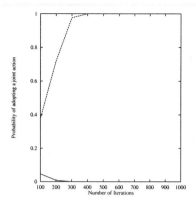

Fig. 1. Game matrix where a_3 and b_3 are individually preferable to the agents, also only $\langle a_3, b_3 \rangle$ is the Nash Equilibrium (left). The probability plots for the joint actions $\langle a_1, b_1 \rangle$ (solid) and $\langle a_3, b_3 \rangle$ are shown on the right.

desirable payoff. This is because the payoff matrix is constructed such that a_3 is the best response (actually in this example, a_3 and b_3 are also the agents' dominant strategies) of agent A irrespective of B's choice and b_3 is the best response of agent B irrespective of A's choice. Our next experiment involved unilateral revelation of information. We first allowed only A to reveal its actions. In this case, B will take its action to maximize its payoff, given that A has taken a specific action. As a_3 and b_3 are the agents' dominant strategies of the agents, they still converge to the "undesirable" solution 2 (left). A similar result is obtained when B reveals its actions 2 (middle). In the final set of experiments, both A and B were allowed to reveal their actions. However, the agents still converged to the nash equilibrium. The agents were unable to overcome the "lure" for short-term profits inspite of the extra information. For example, if agent A reveals that it will take action 1, agent B will start exploiting agent A and take action 3. So, agent A's payoff will reduce to 0 and it will lose trust in agent B and will take action 3 henceforth. Thus, the agents will both take action 3 and converge to the Nash equilibrium 2 (right).

We then reduced A's payoff for $\langle a_3, b_1 \rangle$ and B's payoff for $\langle a_1, b_3 \rangle$ to 9 so that both $\langle a_3, b_3 \rangle$ and $\langle a_1, b_1 \rangle$ are pure Nash Equilibria (figure 3(left)). However, $\langle a_1, b_1 \rangle$ is the desirable solution. The corresponding probability plots are reported in figure 3(right). Here too the EUPs converge to the undesirable Nash Equilibrium and for the same reasons as listed above. The quadratic programming approach [4] produced a mixed strategy (probability distribution) of $[0, 0, 1]$ and $[0, 0, 1]$ for the agents A and B respectively. This corresponds to selecting the $\langle a_3, b_3 \rangle$ action combination. Thus, our EUPs learn almost the same strategy as the mixed-strategy learners seeking Nash Equilibrium. The next set of experiments were aimed at partial revelation of information. When either agent reveals the information about the action taken, the agents learn to overcome their "greedy" strategies and converge to the desirable solution. For example, if

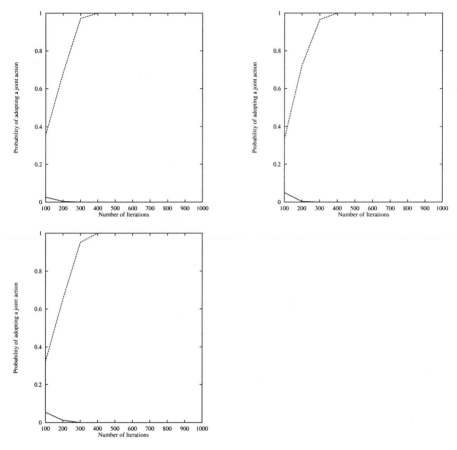

Fig. 2. The probability plots for the joint actions $\langle a_1, b_1 \rangle$ (solid) and $\langle a_3, b_3 \rangle$ are shown when A reveals its actions, B reveals its actions and both(with equal probability) reveals their actions (figures from left to right).

agent A reveals that it is taking action 1, the maximum payoff agent B can get is by taking action 1 too and hence, they both learn to take action 1 as shown in figure 4 (left). A similar argument holds for agent B revealing its information to agent A as shown in figure 4 (middle). When both agents can reveal information (in any iteration) with equal probability they learn that their best response is to select action 1 when the other agent selects action 1 as shown in figure 4(right). Thus, in both bilateral and unilateral information revelation games, convergence to a desirable solution is acheived.

For the probability plot in figure 5 (right), the matrix on left has both $\langle a_1, b_1 \rangle$ and $\langle a_3, b_3 \rangle$ as pure Nash Equilibria. The EUPs learn to adopt the desirable action combination $\langle a_1, b_1 \rangle$ in most runs. A similar result is obtained in both unilateral and bilateral information revelation. The probability plots are shown in figure 6.

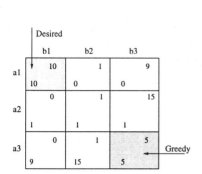
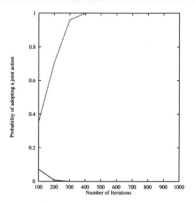

Fig. 3. Game matrix where a_3 and b_3 are relatively preferable to the agents while both $\langle a_1, b_1 \rangle$ and $\langle a_3, b_3 \rangle$ are the Nash Equilibria (left). The probabilitry plots for the joint actions $\langle a_1, b_1 \rangle$ (solid) and $\langle a_3, b_3 \rangle$ are shown on the right.

In the game matrix in figure 7 (left), $\langle a_3, b_3 \rangle$ is the only pure Nash Equilibrium. From figure 7 (right) we can see that the EUPs do not succeed in selecting the desirable solution more often than $\langle a_3, b_3 \rangle$ (the Nash Equilibrium solution).

The profile learned by 1-level mixed strategy agent for the matrix in figure 7 (left) is $[0.09, 0, 0.91]$ and $[0.09, 0, 0.91]$ for A and B respectively. This gives an expected reward of 5.45 to each of the mixed-strategy equilibrium learners, whereas our EUPs receive expected reward of 5.0 for selection of the joint-action $\langle a_1, b_1 \rangle$ alone. Our next set of experiments on this matrix were aimed to understand the effect of unilateral information revelation. Agent B revealed its information (action selection) to the agent A. Agent B's average payoff for selecting action 1 is the highest and hence, nondeterministically, it will tend to select action 1 more often than others. Given this situation, agent A will start taking its best response to agent B's action selection (which B has revealed). Thus, they will converge to the the desirable solution and both take action 1. The probability plots are given in figure 8 (middle).

In the next set of experiments, Agent A revealed its inform ation (action selection) to the agent B. Agent A's average payoff for selecting action 1 is the highest and hence, nondeterministically, it will tend to select action 1 more often than others. Given this situation, agent B will start taking its best response to agent A's action selection (which A has revealed). Agent B will get its highest payoff by selecting action 3 when agent A selects action 1. This will lead to less payoff for agent A. However, the average payoff received by agent A for selecting action 3 is lower than action 1 and hence, it will still choose action 1. Thus, revelation of information by agent A is harmful to it. Its average payoff is less when it reveals its actions to agent B. The probability plots are given in figure 8 (left). We ran an experiment when we increased agent A's average payoff for action 3. Agent A learned to avoid its mistakes and converged to the Nash equilibrium. Finally, we ran an experiment where both the agents had equal

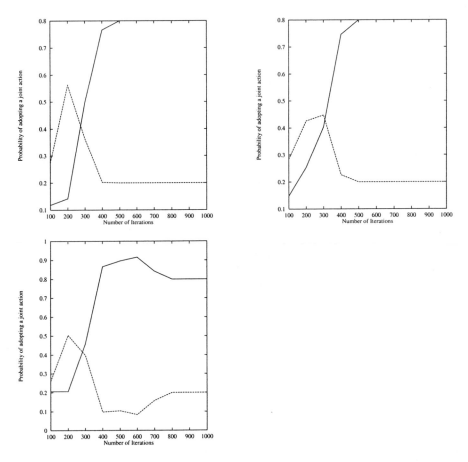

Fig. 4. The probability plots for the joint actions $\langle a_1, b_1 \rangle$ (solid) and $\langle a_3, b_3 \rangle$ are shown when A reveals its actions, B reveals its actions and both(with equal probability) reveals their actions (figures from left to right).

chance of revealing the action taken in any particular iteration. Given that both agents have their best average payoff in selecting action 1, they learn to take action 1 and hence reach the desirable solution.The probability plots are given in figure 8 (right).

The question of mutual trust can be highlighted in the matrix in figure 7 (right). If a combination of $\langle a_1, b_1 \rangle$ is being played, agent B has the incentive to change its action from b_1 to b_3 to increase its payoff from 10 to 11. When it makes such a change, A's optimal response would be to change from a_1 to a_3 to increase its payoff from 4 to 5. Thus, in their haste to respond optimally to the current situation, both agents converge to an equilibrium which pays them half of what they could have got if they had showed restraint. Each of our EUPs, on the other hand, trusts the other's probability-distribution over the actions (given that one of them reveals information about its action selection) and selects its

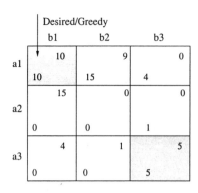
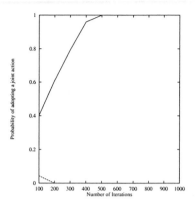

Fig. 5. Game matrix where a_1 and b_1 are relatively preferable to the agents while both $\langle a_3, b_3 \rangle$ and $\langle a_1, b_1 \rangle$ are the Nash Equilibria (left). The probabilitry plots for the joint actions $\langle a_1, b_1 \rangle$ (solid) and $\langle a_3, b_3 \rangle$ are shown on the right.

action stochastically based on that distribution. Thus they progressively tend towards the mutually beneficial part of their search space, emulating restraint which leads to mutual benefit.

Our experimental results suggest that, in a bilateral information revealing scenario, an agent will learn to overcome its greedy (myopic) choice given the following condition. Let us consider two agents A and B. Each agent has n actions to choose from, where a_is are the actions of agent A and b_is those of agent B. Now, let a_x give the maximum expected payoff to agent A. Under this condition, agent A will have a predilection to choose action a_x during the initial exploration phase. Let us consider an iteration where agent A reveals its action to agent B. Let a_x be the chosen action for agent A. Now, agent B will choose its best response to action a_x (it will select the action which gives it the maximum average payoff given A's action). Let this action be b_y. Let R_a be the payoff to agent A due to action-pair selection (a_x, b_y). If R_a is greater than the average payoff due to the other actions that agent A can take ($R_a > \max_{w \in OA} R_w$ where OA represents other actions of agent A), the agents will learn to converge to the desirable action-pair (a_x, b_y).

5   Future Work

Our basic result is that there are certain game-structures, where stochastic modeling agents can converge to high payoff structures which will be missed by more complex modeling learners that are designed to produce Nash Equilibrium [4]. We do not tout our empirical results as an argument for always using EUPs.

However, our observation clearly demonstrates that learning to select a Nash Equilibrium is not necessarily the best an agent can do, and that agents who are not bound by such criteria can sometimes do better. In future, we plan to study

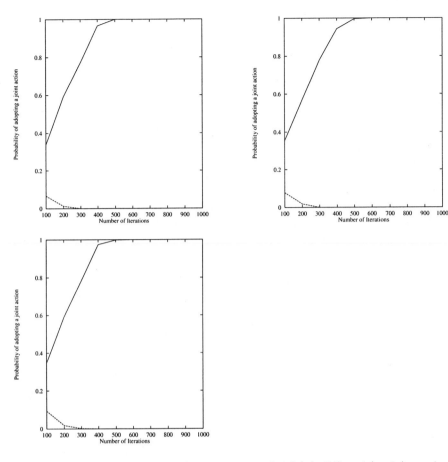

Fig. 6. The probability plots for the joint actions $\langle a_1, b_1 \rangle$ (solid) and $\langle a_3, b_3 \rangle$ are shown when A reveals its actions, B reveals its actions and both(with equal probability) reveals their actions (figures from left to right).

the theoretical basis for selection of a non-equilibrium solution and identify the nature and extent of mutual trust necessary to do so.

An interesting observation from our results is that unilateral or bilateral action revelation can lead to a more trusted behavior resulting in higher payoffs to the agent. We have explained this non-intuitive phenomena by analyzing the difference in evolution of learned behavior with and without the presence of action revelation. Revealing can obviously lead to worst result for the revealer in a number of scenarios, e.g., the Prisoner's Dilemma [6]. We plan to investigate an adaptive revealing strategy that can learn whether or not to reveal choices in a given game.

Acknowledgement

This work has been supported, in part, by an NSF CAREER Award: IIS-9702672.

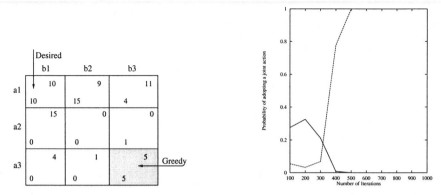

Fig. 7. Game matrix where a_1 and b_1 are relatively preferable to the agents but only $\langle a_3, b_3 \rangle$ is the Nash Equilibrium (left). The probabilitry plots for the joint actions $\langle a_1, b_1 \rangle$ (solid) and $\langle a_3, b_3 \rangle$ are shown on the right.

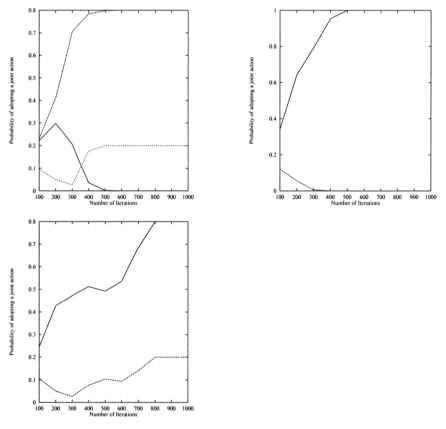

Fig. 8. The probability plots for the joint actions $\langle a_1, b_1 \rangle$ (solid) and $\langle a_3, b_3 \rangle$ are shown when A reveals its actions, B reveals its actions and both(with equal probability) reveals their actions (figures from left to right). The two dotted lines in the figure on the left represent actions $\langle a_1, b_3 \rangle$ and $\langle a_3, b_3 \rangle$ with agents learning $\langle a_1, b_3 \rangle$.

References

1. S. J. Brams. *Theory of Moves*. Cambridge University Press, Cambridge: UK, 1994.
2. C. Claus and C. Boutilier. The dynamics of reinforcement learning in cooperative multiagent systems. In *Proceedings of the Fifteenth National Conference on Artificial Intelligence*, pages 746–752, Menlo Park, CA, 1998. AAAI Press/MIT Press.
3. D. Fudenberg and K. Levine. *The Theory of Learning in Games*. MIT Press, Cambridge, MA, 1998.
4. J. Hu and M. P. Wellman. Multiagent reinforcement learning: Theoretical framework and an algorithm. In J. Shavlik, editor, *Proceedings of the Fifteenth International Conference on Machine Learning (ML'98)*, pages 242–250, San Francisco, CA, 1998. Morgan Kaufmann.
5. M. L. Littman. Markov games as a framework for multi-agent reinforcement learning. In *Proceedings of the Eleventh International Conference on Machine Learning*, pages 157–163, San Mateo, CA, 1994. Morgan Kaufmann.
6. R. D. Luce and H. Raiffa. *Games and Decisions: Introduction and Critical Survey*. Dover, New York, NY, 1957.
7. O. L. Mangasarian and H. Stone. Two-person nonzero-sum games and quadratic programming. *Journal of Mathematical Analysis and Applications*, 9:348–355, 1964.
8. M. Mundhe and S. Sen. Evaluating concurrent reinforcement learners. Proceedings of the International Conference on Multiagent Systems, 2000.
9. J. F. Nash. Non-cooperative games. *Annals of Mathematics*, 54:286–295, 1951.
10. T. Sandholm and R.H.Crites. Multiagent reinforcement learning in the iterated prisoner's dilemma. *Biosystems*, 37:147–166, 1995.
11. T. W. Sandholm and R. H. Crites. Multiagent reinforcement learning and iterated prisoner's dilemma. *Biosystems Journal*, 37:147–166, 1995.
12. S. Sen, M. Sekaran, and J. Hale. Learning to coordinate without sharing information. In *National Conference on Artificial Intelligence*, pages 426–431, Menlo Park, CA, 1994. AAAI Press/MIT Press. (Also published in *Readings in Agents*, Michael N. Huhns and Munindar Singh (Editors), pages 509–514, Morgan Kaufmann Publishers Inc., San Francisco, CA, 1998.).
13. R. S. Sutton and A. G. Barto. *Reinforcement Learning: An Introduction*. MIT Press, Cambridge, MA, 1998.
14. C. J. C. H. Watkins and P. D. Dayan. Q-learning. *Machine Learning*, 3:279–292, 1992.
15. G. Weiß. Learning to coordinate actions in multi-agent systems. In *Proceedings of the International Joint Conference on Artificial Intelligence*, pages 311–316, August 1993.

Distributed Trust in Open Multi-agent Systems

Yosi Mass and Onn Shehory

IBM Research Lab in Haifa, The Tel Aviv Site
2 Weizmann St., Tel Aviv 61336, Israel
{yosi,onn}@il.ibm.com

Abstract. Facilitated by the rapid growth of the Internet, electronic commerce is growing exponentially. As a result, millions of players participate in electronic trade, yet many of these players are strangers to each other. This implies mistrust, which may bring about manipulative and malicious trade behaviors among the parties. This problem intensifies in electronic environments where agents act on behalf of humans. There, self-interested, utility-maximizing agents, have a strong motivation, and no moral means against, malicious action. Attempts to prevent such misbehavior usually concentrate on designing non-manipulable mechanisms. Yet, these tend to be either computationally intractable or sub-optimal. We suggest a new approach: a mechanism that allows agents in an open system to establish trust among themselves and to dynamically update this trust. Although we rely on certificates for our solution, we do not require (in contrast to previous solutions) any centralized certificate authority system, nor do we require some well known, trusted parties. Our solution is fully distributed, it is computationally feasible, and can be easily added to any agent architecture.

1 Introduction

Trust is a major vehicle for enabling trade. The need for trust establishment increases in societies of trading electronic agents. This necessity intensifies as the size of the trading society grows larger and electronic markets become open to electronic strangers. As suggested e.g., in [12], trust can be established through the presentation of third party certificates. Yet, previous work assumes that: (1) there exists a publicly known and trusted third party reputation systems [14]). However, the requirement that an agent reveals its identity to participate in trade activity [12]; (2) to establish trust, there is a need to know the agents with whom one interacts (e.g., in multi-agent violates the agent's right for privacy (of itself and of its sender). In agreement with others, we believe that for trade to take place among electronic entities, it is necessary to establish trust among them. Unlike previous work on agent trust, though in agreement with [5], we adopt the approach according to which identity disclosure is not required for trust establishment. Furthermore, there is no need for a centralized certification mechanism or globally known trusted third parties. In this sense, our work is similar to multi-agent referral systems (e.g., [13]). Yet, in contrast to such work, we use certification and, when trusted parties exist, they can be part of the

R. Falcone, M. Singh, and Y.-H. Tan (Eds.): Trust in Cyber-societies, LNAI 2246, pp. 159–173, 2001.
© Springer-Verlag Berlin Heidelberg 2001

solution. The validity of the approach we take was shown in [5]. We base the architecture presented in this paper on the mechanism developed and presented there. We add upon that work the following: (1) an agent architecture which supports trust and dynamic update of trust level based on new inputs and historical encounters; (2) means for agents to verify capabilities claimed by other agents.

Our solution (as in [5]) is based on the use of certificates. A certificate is an electronic document that is signed by some *issuer* and contains some *claims* about a *subject*[1]. For example a certificate may establish the fact that a user (the subject) is an employee (a claim) of a company (the issuer). The subject is represented by a *public key* and it should hold the *private key* that corresponds to that public key to prove that it is really the subject. The proof is done by the subject signing with its private key on some challenge sent to it by the verifier and then the verifier can check the signature using the public key that is in the certificate. A typical certificate contains no claims and only a subject name and is used for binding the subject's public key to the subject's name. This is sometimes called an Identity certificate and it is used for authenticating a user to a system and proving its identity. When dealing with agents on the internet and assuming that an agent may be a subject in some certificates, there is no meaning to the agent's name but what we are interested in are the claims about the agent which is represented by the subject's public key.

The mechanism in [5] enables trade parties, among them agents as well, to define policies for mapping strangers to predefined business roles. The level of trust of an agent Y regarding a stranger agent X, and the corresponding level of access to Y's resources and services X is granted, is based on Y's policy regarding access provision to various roles. The process of trust establishment – role assignment and access provision – is performed in a fully distributed manner, where any party or agent may be a certificate issuer, and it is not required that certificate issuers be known in advance. Instead, it is sufficient that, when requested, an agent that issues certificates provides sufficient certificates from other issuers to be considered a trusted authority according to the policy of the requesting party. This allows distributed trust build-up among electronic market players, and in particular among agents.

Our approach is based on the automated role assignment presented in [5] and borrows from previous role-based access control mechanisms, however it extends those to open multi-agent systems. Access control mechanisms usually use the identities of subjects to map them to roles, based on a static table. The solution we adopt [5] is different – it does not require identities for the assignment of roles. When an agent, X, approaches another agent Y and requests access to a service or a resource held by Y, agent Y can, using that mechanism, map X to a role. Y will perform this mapping based on X's certificates, on a given role-assignment policy set by Y in advance (or by any other owner of the resource, if Y is not the owner), and on the roles of the agents who issued the certificates presented by

---

[1] Details regarding public keys, private keys, digital signatures and certificates are found in Appendix B and in multiple web pages, e.g. [1,2,3].

X. The role is then fed as an input to the traditional role-based access control mechanism (as described in proceeding sections). Based on this, we developed a trust establishment architecture for open multi-agent systems. We demonstrate how this architecture fits into standard agent architectures. Further, our solution supports capability verification, which is a missing service in open MAS.

1.1 Potential Applications

One of the fast growing Internet applications is the domain of electronic market-places (eMP). These serve as meeting places for buyers and sellers and enable Business-to-Business (B2B) transactions. We believe that the proliferation of eMPs (there are currently hundreds of B2B eMPs) will continue rapidly. In particular, the number of vertical eMPs that specialize in a specific business area (e.g., eMP for electronic parts, for cars, paper etc) will grow. This trend should allow a company or an individual to be registered in many eMPs. Agents may play an important role in such environments in locating relevant businesses and performing negotiations on behalf of their users across markets. In such open environments, agents can locate other agents using standard agent location mechanisms (e.g., [7, 8]). However, to perform meaningful business interaction with them, some level of trust must be established between the agents.

For instance, an agent's policy may be to interact only with agents that were developed by a "good company" (we assume that an agent has a certificate signed by its developer). A "good company" can be either known locally to the agent, or certified by at least two already known "good companies" (we assume that a company can recommend another company as a developer of good agents). In this example there is no need for an agent to know all of the companies which are developers of good agents. It would suffice that the agent knows a small subset of these developers, and the latter can provide recommendations regarding other companies.

2 Trust Establishment between Agents

Agents in MAS need to establish some level of trust in each other prior to service exchange interaction. We differentiate between two types of agent interactions

- locating a trusted enough agent to deal with
- authorizing requests to accessing agents

The first problem of locating a trusted agent can be viewed as an extension of a MatchMaker as suggested in the RETSINA system [12]. The RETSINA system assumes that each agent has a unique identifier and a certificate from some central Agent Certification authority (ACA) that binds its public key to its identifier such that the agent can prove that it owns the private key that corresponds to the public key in the certificate. The RETSINA system also assumes that an agent can register with the Matchmaker claiming to have some capabilities. RETSINA has no explicit assumption regarding capability verification, however no mechanism is provided for the MatchMaker to verify that the registering agent has its claimed capabilities. Agent X that needs to locate agents with requested capabilities can query the MatchMaker for certain capabilities and the MatchMaker returns the list of matched agents with their certificates. Then agent X can communicate with the agents and authenticate them through their certificate.

There are two deficiencies in the above solution. The first is that it assumes a central ACA that certifies all of the agents. This is not scalable to the whole community of agents that operate on the Internet. The second deficiency is that it is not clear how the agents' claimed capabilities can be verified, either by the MatchMaker during the registration process or later by a requesting agent. We suggest another architecture where there is no need for a central ACA, and instead each agent X may have one or more certificates certifying its capabilities and its performance. One such type of a certificate can be issued by X's developer or deployer (e.g., the agent was written by IBM). Other certificates can be issued by other third parties that have previously used X's services and can provide recommendations about its performance and trustworthiness. The solution requires an infrastructure for various types of certificates, which carry some information about the certificate holder. The Trust Establishment work in [5] suggests such a solution.

An agent X that needs to locate a trusted agent with appropriate capabilities can define a trust policy as in [5]. Then when the MatchMaker returns the list of matched agents, their certificates can be fed into agent X's trust policy. The policy can say which of the returned agents is trusted enough according to the policy rules. For example, a policy rule may state the following: if an agent was developed by a "trusted" company, then the agent can be trusted. The policy can also define rules regarding who is a trusted developer and the Trust Establishment mechanism can collect more certificates about the developer to learn whether it is trusted according to the policy.

The second problem of authorizing accessing agents is a sub-problem of the Access Control (AC) problem. Specifically, we are interested in the following problem. An agent X sends a request to agent Y; agent X is not known to Y in advance; should agent Y respect X's request? The question of what services should Y provide to X can be mapped to the AC problem of deciding what resources should Y provide X access to. Our approach to trust establishment among agents is based on providing each agent, as part of its internal architecture, with an access control component, which allows the agent to make such decisions. The access control component answers queries on whether or not an agent X may perform action A on resource R.

In a typical access control systems the requesting agent X is identified to the provider agent Y by a certificate that includes the identity of X. The certificate is used to identify the requesting agent. Public key certificates are data structures that bind public key values to entities (agents in our case). The binding is achieved by having a trusted Certification Authority (CA) digitally sign each certificate.

Note that the identity of X must either be known to Y, or Y may approach an Agent Name Server (ANS) [10] to verify the identity, however this verification can only assure Y that X exists at a given IP address. A multi-agent trust establishment approach similar to the typical AC process was presented in [12]. The typical AC process has several drawbacks, as follows:

1. It requires that agents either be known in advance to each other, or use some ANS.
2. It requires that each service provider agent, (which is virtually any agent) hold an exhaustive, static mapping of all of the agents that may request its services to access permissions.
3. It violates agents' privacy (i.e., its anonymity is lost).

Requirements 1 and 2 can only hold in closed MAS, and are unrealistic in open electronic markets. Privacy may also be an important service that should be enhanced by appropriate interaction mechanisms. To solve this problem we can use again the Trust Establishment [5] solution to map an accessing agent based on its certificates to a role and use that role in an access control component that will be part of each agent.

For instance, suppose that there are several certificate issuers. A rule on a field of the certificate can be, e.g., to allow access to all agents presenting a certificate signed by a specific issuer, e.g., IBM. This approach adds flexibility to the certificate role assignment.

The TE module enables an agent (or its owner) to define a flexible policy for role assignment. This supports: dynamic ad-hoc inter-agent relationships; evolvement of a web of trust among agents; separation of authority between the agents (and their organizations) issuing certificates and the agents defining and using the policy to control access to their services and resources.

One may observe that the approach we take may lead to an iterative certificate request process. An agent Y who was approached by agent X may, to establish trust with X, request certificates from X's issuers, then from their issuers, and so on. This may result in long certificate chains, and consequently increase communication loads. Yet in practice, relying on the analysis of small world networks [11], the depth of search will be very small [2].

3   Related Work

As stated in previous sections, the use of public key certificates for trust establishment among agents was previously suggested within the RETSINA framework [12]. Yet, in that work, identity disclosure was required, the presence of a certification authority was needed, and no flexible AC policy was supported. Other shortcomings of that solution that our solution overcomes were detailed in Section 2.

---

[2] In a network where each entity knows, on average, dozens of other entities, the average depth of search from one entity to an unknown one will be small (e.g., depth < 10 for 10^9 entities).

A "web of trust" model for public key certificates was first deployed in the Pretty Good Privacy (PGP, [15]) secure e-mail system. However, since e-mail systems do not exhibit many MAS properties, PGP does not provide an architecture to be used in agents. In addition, its policy support is very limited comparing to [5] on which we rely.

IETF's Simple Public Key Infrastructure (SPKI) [2] suggests trust management embedded in certificates. There, identity certificates require a global name space with unique names such as the X.500 DN. This is problematic in open MAS. As shown e.g., in [5], there is no need for identity binding.

There are several reputation mechanisms (e.g., [14]) that were suggested to enhance trust among electronic business parties, however those mechanisms do not take advantage of PKI technology as we do.

4   The Architecture

Our solution is based on an architecture we developed to address the unique trust establishment needs of agents in a large open MAS. According to this architecture, agents may have, in addition to their individual expertise, three major behaviors: certificate issuers, service requesters, and service providers. Note that the behavior-based architecture fully complies with leading multi-agent architectures (e.g., [10]) and therefore can be easily implemented as part of them. In our solution, each agent may implement any combination of these three behaviors. For instance, an agent may request some services from other agents, behaving as a service requester and, possibly concurrently, provide other agents with different services, behaving as a service provider. Yet, when an agent X is a service requester, it usually needs to carry with itself some certificates, and these will be provided by agents (or other entities) that are behaving as certificate issuers and are willing to certify X. Their willingness to provide X with certificates may be based on the trust they have in X or on some cross-certification policy. Note that a service provider agent Y is not required to respect X's certificates, and Y's decisions will be based on its role assignment policy. Yet, unlike traditional role assignment policies, we introduce a strategic role assignment policy (in a consequent section). It may happen that Y will ask X's issuers to present certificates, so that Y can establish trust with X's issuers, to establish trust with X as well. The multi-agent architecture is depicted in Figure 1. Note that organizations and individuals that the agents may be representing do not appear in the Figure, although they may take part in setting policies for both, certificate issuance (for the issuer behavior) and for role assignment (for the provider behavior). In addition, organizations may directly issue certificates, as assumed in [5].

The multi-agent architecture is not complete without details of the internal agent architecture that are necessary for implementing the three behaviors presented above. In particular, it is necessary to provide details regarding the architecture of an agent that implements the provider behavior, since this behavior plays the most important role in the trust establishment architecture. According

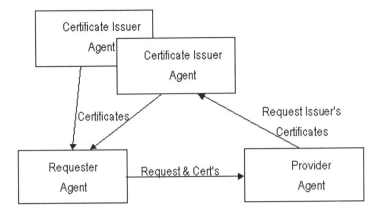

Fig. 1. A multi-agent architecture for trust establishment.

to our architecture, each agent that may implement a service provider behavior has an embedded trust establishment architecture. This architecture consists of a role assignment module, an AC mechanism and a policy definition tool (see Figure 2).

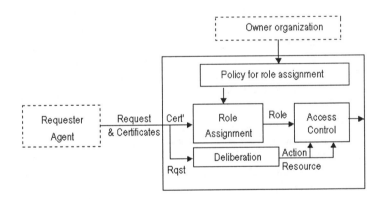

Fig. 2. The internal agent architecture for the provider behavior.

The embedded trust establishment architecture functions as follows. When a requester agent X sends a request to a provider agent Y, it attaches certificates to its request. Y separates between the request and the certificates: the certificates are sent to the role assignment module and the request is sent to the deliberation module. The role assignment module retrieves role assignment policies and, according to X's certificates and the policy, assigns X a role. Concurrently, the deliberation module analyzes X's request to find the resources and actions needed for its fulfillment. The role, resources and actions are then fed

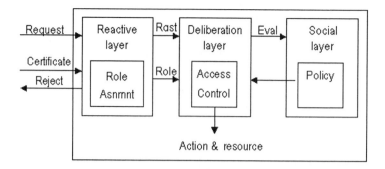

Fig. 3. Incorporating the trust architecture in a layered architecture.

into the AC mechanism, which in turn provides a decision on whether or not X's request should be respected. Note that policies can either be inserted and edited by the agent's owner, or can be dynamically updated via an internal evaluation and learning mechanism. Dynamic policy update is a unique feature of our solution and is most adequate for agents, who may take advantage of their ability to learn from previous encounters.

The architecture we describe above can easily fit into common agent architectures. One of the most common architectures is the layered architecture. We demonstrate the incorporation of the trust establishment architecture in a layered architecture that consists of three layers: reactive, deliberative and social. As depicted in Figure 3, a request and a certificate are first received by the reactive layer. Certificates are checked and, when appropriate, a rejection is performed immediately by the reactive layer. In other cases, the reactive layer assigns a role to the certificate holder and passes the role and the request to the deliberative layer for reasoning, planning, and access control. The deliberation layer plans and finds what resources are necessary for the request, and feeds this information into the access control component. Based on this and the trust policy, access is provided and the agent proceeds with action execution. The deliberation layer collects execution results and passes a success evaluation to the social layer. The social layer records interactions with external entities and learns from these for future interaction. In particular, it updates the trust policy, which feeds into the deliberation layer in future encounters.

5   Strategic Role Assignment

As stated above, we follow the policy-based role assignment approach of [5]. However, our solution adds a unique feature: instead of static decision-making, we take advantage of the fact that an agent architecture (in the case of deliberative agents) usually consists of components that can reason and plan, before proceeding to execution (e.g., [6]). As the decision upon whether or not an agent X should be provided by Y with a service is based on external-environment and internal-agent parameters, it is necessary to take such parameters into consideration. In particular, it is possible for a self-interested, rational agent to behave strategically and take near-optimal decisions. An agent Y is given its goals, a set of alternative activities (among them – the provision of services to requesting agents), the value associated with each activity, and the associated risk (in terms of success/failure probability). To dynamically plan for the best set of alternatives, Y can follow the Hierarchical Task Network (HTN) approach [4], implementing the extensions suggested in [6]. In particular, during execution of some activities, some of the parameters may change. For instance, suppose Y has provided a service to X, and that X was initially considered a trusted business party (based on the certificates X presented and the policy of Y's), which means that the risk level of doing business with X was low. At that time, additional service requests from X were highly prioritized in Y's plan, as they promised value with little risk. Suppose now that X did not meet Y's trust expectations during their business interaction (e.g., X failed to pay for the service), however no issuer has changed or revoked X's certificates. In Y's view, though, X's risk level has changed (i.e., X is less trustworthy). Y can use its parameter estimators, embedded into its extended HTN (as in [6]), to dynamically re-assess the value of each activity in which X is involved. Based on this re-assessment, Y can also change its trust establishment policy. For instance, it may downgrade the trust level afforded to agents who present certificates from X (when the identity of X is known). In addition it may downgrade the trust level afforded to agents who present certificates from X's issuers. Thus, the use of dynamic planning and action evaluation enables agents to modify their trust establishment policy to best fit the changes in their business environment and improve their overall expected gain from inter-agent activity.

Note that an alternative to using HTNs may be, e.g., using decision trees, where the exponentially large search space (which is a well-known problem in planning) is pruned using some AI pruning heuristics. Yet, such mechanisms do not provide solutions to interleaving planning and execution, nor do they explicitly provide means for re-evaluation of alternatives (as [6] does). The latter are necessary for agents deployed in an open dynamic environment.

6   Certificates

The architecture presented above is based on the trust establishment (TE) infrastructure embedded in the agents, which allows agents as well as their owners to define flexible policies, based on a "web of trust relationships" supported by public-key certificates from trusted agents. In other solutions, certificates are used for binding agents to a pubic key (Identity certificate). Our solution relies on certificates with extension attributes, hence we extend the use of certificates in MAS to not only identifying[3] an agent, but also to claim something about the agent that holds the certificate. We make an intensive use of the certificate's

---

[3] In our approach, agent identification is required only by means of its public key. Other identifying attributes may enhance trust but are nevertheless optional.

extensions and the policy is based on rules with constraints on those extensions. Our solution is compatible with IETF's PKI [1] and is based on X509v3 [1] certificates, however it is not limited to X509v3 certificates and can work with any certificate format that supports extension attributes. As defined in [5], a Trust Establishment certificate is an object of the following type:

$$C = Cert(Issuer,\ Subject,\ [field_1,\ value_1],\ ....,\ [field_k,\ value_k]). \tag{1}$$

A certificate of this type expresses assertions in a trusted and non-repudiated manner and enhances exchange of trust-related information between unrelated agents and other parties in the web-of-trust. A certificate is signed by the *Issuer*, which implies that the issuer certifies that the following assertions hold regarding the *Subject*: $field_1 = value_1$, $field_2 = value_2$, ... , $field_k = value_k$. Note that both *Issuer* and *Subject* should include a unique identifier of the issuer and the subject, respectively. Such an identifier does not necessarily include an explicit identity such as a name or an IP address. However, the unique identifier must either include the public key of the entity it identifies or provide a unique link or mapping to this key. An example of a developer's certificate may be:

$$C1 = Cert\ (IBM,\ Agent\ X,\ certType = Developer) \tag{2}$$

which states that Agent X was developed by IBM. Note that when we write "Agent X" we mean the entity that owns the private key that corresponds to the public key which is in the certificate.

If agents are, for instance, involved in stock portfolio management, an example certificates may be:

$$C2 = Cert\ (AmEx,\ Agent\ X,\ certType = Recommendation,\ Value = 9) \tag{3}$$

which states that AmEx recommends Agent X with recommendation value equal to 9, or:

$$C3 = Cert\ (CityBank,\ Agent\ X,\ certType = Broker, \\ Type = Nasdaq,\ Service = good) \tag{4}$$

which states that CityBank asserts that Agent X is a stock broker for stocks traded on Nasdaq, with a good service quality.

A role assignment policy of an agent may require several such certificates to establish trust with a requesting agent. For instance, the policy may state that an agent should be granted access to stock trading services if it holds two certificates from recognized banking institutes that assert that the agent that holds them is a good stock broker. Here the policy does not require that the identity of the requesting agent be explicitly disclosed – it is sufficient to have certificates that do not include other identity attributes of the holder. Further technical details regarding TE certificates are in Appendix A.

Note that in contrast to previous use of public-key certificates for agent trust [12], where certificates are viewed as an (only) binding between a public key and an identity, trust establishment certificates are more expressive. Since they must express a wide spectrum of statements, their format cannot be predefined and stringent. As a result, certificates can be of many different types, composed of different fields that have different meanings. These differences may impose interoperability problems in open environments, yet [5] provides some means to overcome such problems. In addition, interoperability obstacles, that results from different typing and meanings can be addressed using solutions such as LARKS [9].

7   Conclusion

We have presented an application of distributed trust establishment to open multi-agent systems. Note that although this work is based, in part, on previous work [5], the attributes of open MAS imposed a need for a unique architecture in which the trust establishment modules are embedded. We introduce this architecture, describe the way in which it functions, provide details regarding its internals and demonstrate how it fits into common agent architectures. Our work further extends [5] in the introduction of dynamic policy update. It also provides a means for agents to verify claimed capabilities of other agents. Though, the major contribution of this work is in allowing designers of open MAS to avoid the burden of devising very complex mechanisms to avoid malicious behavior of commerce partners on the Internet. Instead, we propose an architecture, backed up by the necessary mechanisms, for establishing trust among agents in a completely distributed manner, with no central CA, no need to know agents in advance, and actually no need to know the other agents at all. The solution we propose complies with up to date Internet standards, hence implementation and deployment are rather simple.

References

1. Internet X509 Public Key Infrastructure,
   http://www.ietf.org/ids.by.wg/pkix.html.
2. Simple Public Key Infrastructure (SPKI),
   ftp://ftp.ietf.org/internet-drafts/draft-ietf-spki-cert-theory-02.txt.
3. http://developer.netscape.com/docs/manuals/security/pkin/contents.htm.
4. Erol, K., Hendler, J., Nau, D.: HTN Planning: Complexity and Expressivity. In Proceeding of the 12th National Conference on Artificial Intelligence (AAAI-94), Seattle, 1994.
5. Herzberg, A., Mass, Y., Mihaeli, J., Naor, D., Ravid, Y.: Access Control Meets Public Key Infrastructure, or: Assigning Roles to Strangers. 2000 IEEE Symposium on Security and Privacy, Oakland, May 2000.
6. Paolucci, M., Shehory, O., Sycara, K.: Interleaving Planning and Execution in a Multiagent Team Planning Environment. Technical Report CMU-RI-TR-00-01. Robotics Institute, Carnegie Mellon University, 2000.
7. Shehory, O.: A Scalable Agent Location Mechanism, Lecture Notes in Artificial Intelligence, Vol. 1757, Intelligent Agents VI, Springer, pp. 162-172, 1999.
8. Sycara, K., Decker, K., Williamson, M.: Middle-Agents for the Internet, Proceedings of the 15th Joint Conference on Artificial Intelligence (IJCAI-97), 1997.

9. Sycara, K., Lu, J., Klusch, M.: Interoperability among Heterogeneous Software Agents on the Internet. Technical Report CMU-CS-92-131, Robotics Institute, Carnegie Mellon University, 1998.

10. Sycara, K., Pannu, A., Williamson, M., Zeng, D., Decker, K.: Distributed Intelligent Agents. IEEE expert pp. 36-45, December 1996.

11. Watts, D., Strogatz, S.: Collective dynamics of small world networks. Nature 393, pp. 440-442, June 1998.

12. Wong, H.C., Sycara, K.: Adding Security and Trust to Multi-Agent Systems, Proceedings of Autonomous Agents 99' workshop on Deception, Fraud and Trust in Agent Societies, May 1999, pp. 149-161.

13. Yu, B., Singh, M.: A Multiagent, Referral System for Expertise Location. In Proceedings of the HICSS-32 minitrack on Electronic Commerce, 1999.

14. Zacharia, G., Moukas, A., Maes, P.: Collaborative Reputation Mechanisms in Electronic Marketplaces. AAAI Workshop on Intelligent Information Systems, 1999.

15. Zimmerman, P.: "The Official PGP User's Guide", MIT Press, Cambridge, 1995.

Appendix A: TE Certificates

The Certificate Profiling mechanism and the supporting language presented in [5] enhance the flexibility achieved in trust establishment in decentralized open MAS. We briefly present these below. The mechanism allows different and independent certificate issuers and role assignment modules to negotiate, exchange and understand certificates. According to that mechanism, every certificate must be an instance of a predefined *certificate type*, where the set of all certificate types can be dynamically extended. The *certificate type* must be unique. Hence, it is either registered through some central organization or becomes unique by adding to it the originator prefix (e.g. www.inm.com.employee).

Each certificate type has a *certificate profile*, which defines the certificate structure, namely which fields it is composed of and what types of values can each field admit. The certificate profile must address two issues:

- Syntax - A listing of all fields, types of values each field can admit and the mandatory fields. The syntax is expressed in XML so it is parseable both by certificate producers and by certificate consumer such as the TE modules presented in Section 4.
- Semantics - An explanation of the meaning of each field and its corresponding values. For human users, this can be done using free text, however for agents a careful XML DTD design is necessary.

A TEcertificate object has the following mandatory fields: Issuer, Subject, certType, version, profileURL, issuerCertRepository, subjectCertRepository. Each certificate may includes more fields where a field is identified by its name (a string), and its value can be numeric, string, a range of values, or a set of strings or numbers (some simple examples are in section 6).

The Policy Language and the Role Assignment Module

The TE policy used for mapping *entities* (or agents, in our case) to *roles*, and this is done via logical rules. The language was defined using XML where *roles* are defined at the top level and under each role there are *rules* for *role* membership. We provide below some details of one of two TE policy languages, called **DTPL** (Definite Trust Policy Language), which is monotonic and does not include negative rules. The other one, **TPL** (Trust Policy Language) which includes negative rules is not described here.

A role in DTPL is a group of entities that can represent a specific organizational unit (e.g. contractees, managers, contractors). An agent (or any other entity) in such a group is identified by its public key. A special role is 'self' which includes the key of the policy owner. Each role has one or more rules defining how a certificate holder can become a member in the role. The rules are OR(ed), namely, it is sufficient that one rule holds for mapping an agent to a role. Note that the terms role and group are used interchangeably.

A rule has two parts: 1) a list of required certificates in which the agent to check is the subject and the issuers should belong to some policy role; 2) a function on the certificate fields. An example of a rule appears below. Here, "group 3" has a rule according to which an agent X can be mapped to the group if the following holds:

1. there exists a certificate where X is the subject and whose issuer is in group 2;
2. there exist a certificate where X is the subject and whose issuer is in "group 1";
3. the function defined on the fields of the certificates holds. The function itself is expressed as constraints on the attributes in the two relevant certificates.

The rule is written as follows:

```
<GROUP NAME="Group 3">
 <RULE>
  <INCLUSION ID="C1" TYPE="T1" FROM="Group 2"></INCLUSION>
  <INCLUSION ID="C2" TYPE="T2" FROM="Group 1"></INCLUSION>
 </RULE>
</GROUP>
```

Note that DTPL is more expressive than in the example above. It includes several tags, allows loops, and supports code import.

A Mapping Proof

As mentioned in Section 4, policies are fed into a *role assignment module*, which receives certificates and maps agents who are subjects of the certificates to roles according to the policy. The mapping to roles requires only limited information about the certificate issuers and in particular knowing the role to which the issuers are mapped and not their identity. For instance, suppose the role assignment module is unable to decide whether X is in "group 3" because Y, the issuer one of X's certificates, is not known to be in "group 2". The role assignment module can attempt to map Y to group 2, and if it succeeds, it can in turn map X to group 3.

When a member X is successfully mapped to a role G, a proof (X, G) of the mapping is kept as part of the mapping information. proof (X,G) is defined as the set of certificates that were used in the "Inclusion" statements in the rule that was used for the mapping. Since certificates can expire or be revoked, the mapping is dependent on the validity of the proof. When a new mapping is checked and the mapping algorithm needs to rely on the fact that X' is in G', the certificates in proof (X', G') are verified to be valid. The verification is traced back until the root of the tree.

The depth of a proof is the length of the longest path along the proof until it reaches the "self" group. The policy owner can limit the depth of a proof. The syntax to do it is by adding a "depth" attribute to an "Inclusion" statement.

Appendix B: Public Key Infrastructure

Encryption is the process of transforming information so it is unintelligible to anyone but the intended recipient. Decryption is the process of transforming encrypted information so that it is intelligible again. A cryptographic algorithm, also called a cipher, is a mathematical function used for encryption or decryption. In most cases, two related functions are employed, one for encryption and the other for decryption.

Cryptographic algorithms are usually widely known. Hence, encrypted information is kept secret based a on the use of a *key* (which is a number) with the cipher. Decryption with the correct key is simple. Decryption without the correct key is very difficult, usually computationally infeasible.

There are two major types of key-based cryptography - symmetric and asymmetric. A Public Key Infrastructure (PKI) makes use of asymmetric key cryptography. An asymmetric key pair consists of two keys, a *private key* and a *public key*. The private key should be known only to its owner, whereas the public key may be accessible to anyone. The keys are used for encoding and decoding as shown below (Fig. 4). A message encoded with the public key can only be decoded using the matching private key. A message encoded with the private key can only be decoded with the public key.

Fig. 4. Asymmetric key cryptography.

One use of PKI is for transmission of secret information. To send encrypted data to an entity X, entity Y encrypts the data with X's public key, and X decrypts it with its private key. Since any entity has access to X's public key, anyone can send to X encrypted data. Yet, X is the only entity that knows its private key, hence the only one that can easily decrypt the encrypted data.

Another use of PKI is for digital signatures. Signing is performed by X encrypting some data (usually after it was hashed) with its private key, and then sending it to another entity. Since any entity holds the public key of X's, anyone can decrypt A's message. But only if the message was actually encrypted by X's private key will X's public key allow its decryption. Thus, a successful decryption is a proof that the message was indeed signed by X.

Using signatures, certificates can be formed. A certificate is a claim regarding an entity X, signed by another entity Y. Details regarding certificates are provided in preceding sections of this paper. Commonly, certificates are issued by Certification Authorities (CAs), which are entities that validate identities and issue certificates. CAs may be independent third parties or organizations running their own certificate-issuing procedure. Nevertheless, as we suggest in this paper, any party can issue certificates.

The reader may visit the WWW, where further details on PKI can be found, e.g., at http://developer.netscape.com/docs/manuals/security/pkin/contents.htm.

Modelling Trust for System Design
Using the *i** Strategic Actors Framework

Eric Yu and Lin Liu

Faculty of Information Studies, University of Toronto
Toronto, Ontario, Canada M5S 3G6
{yu, liu}@fis.utoronto.ca

Abstract. The *i**framework was developed to support requirement analysis and high-level design in an agent-oriented system development paradigm. It models intentional dependency relationships among strategic actors and their rationales. As actors depend on each other for goals to be achieved, tasks to be performed, and resources to be furnished, the trust relationships among these actors need to be considered to reason about the opportunities and vulnerabilities these dependencies bring. The concept of softgoal is used to model quality attributes for which there are no a priori, clear-cut criteria for satisfaction, but are judged by actors as being sufficiently met («satisficed») on a case-by-case basis. In this paper, trustworthiness is treated as a softgoal to be satisficed from the viewpoint of each stakeholder. Contributions to trustworthiness are considered using a qualitative reasoning approach. Examples from the smart card domain are used to illustrate.

1 Introduction

Trust is becoming a central issue in today's increasingly networked information systems. For example, in electronic commerce, exchanges often take place among parties unfamiliar to each other, and whose identities may be transitory. Different parts of networks and systems may be operated by parties with conflicting interests or even malicious intent. Furthermore, many technologies, as well as business models, are new and their viability unproven.

Techniques for system analysis and design have, in the past, been focused primarily on addressing functional requirements, assuming that all parties are trusted. Given today's environments, there is need for new techniques that would bring issues of trust, risk, and vulnerability prominently into the system analysis and design process.

The *i** framework [23] was developed for modelling intentional relationships among strategic actors. Actors have freedom of action, but operate within a network of social relationships. Specifically, they depend on each other for goals to be achieved, tasks to be performed, and resources to be furnished. These dependencies are intentional because they are based on underlying concepts such as goal, ability, commitment, and belief. Actors in *i** are strategic because they evaluate their social relationships in terms of opportunities that they offer, and vulnerabilities that they may bring. Strategic actors seek to protect or further their interests. Compared to conventional modelling techniques such as data flow diagramming and object-oriented analysis (e.g., UML), *i** provides a higher level of modelling such that one can reason about opportunities and vulnerabilities. The

R. Falcone, M. Singh, and Y.-H. Tan (Eds.): Trust in Cyber-societies, LNAI 2246, pp. 175–194, 2001.
© Springer-Verlag Berlin Heidelberg 2001

framework has been elaborated in the context of requirements engineering [24], business processing reengineering [27] [25], and software processes [26]. The framework is being extended to form the basis of an agent-oriented system development paradigm.

In this paper, we explore the use of $i*$ for modelling trust and related concepts in system design. Trustworthiness is treated as a softgoal to be satisficed from the viewpoint of each stakeholder depending on others. The concept of softgoal is used to model quality attributes for which there are no *a priori*, clear-cut criteria for satisfaction, but are judged by actors as being sufficiently met ("satisficed") on a case-by-case basis. Contributions to trustworthiness are systematically elaborated and analyzed using a qualitative reasoning approach. The softgoal concept in $i*$ arose from an approach to dealing with non-functional requirements in software engineering. Non-functional quality requirements concern how well the functions to be provided by the system are accomplished, e.g., how speedily (performance), how cheaply (costs), how accurately, etc. While many non-functional requirements are hard to quantify or characterize, e.g., flexibility, maintainability, evolvability, scalability, etc. they are nevertheless essential for the success of a system. An important feature of these non-functional qualities is that they interact with each other in complex ways. The NFR framework [6] [7] offers a graphical notation and framework for systematically elaborating and analyzing the contribution relationships in a network of softgoals. Contributions can be positive and negative, and may be considered partial or sufficient towards addressing some softgoal. The $i*$ framework interleaves non-functional analysis with the functional analysis of system operation. These are done within a network of social actors. Actors may be further differentiated into agents, roles, and positions.

Treating trust as a non-functional quality requirement alongside others during design allows interactions among them to be examined. This includes closely related concepts such as security and risk, as well as more general design issues such as cost, usability, etc. The use of a qualitative reasoning approach allows diverse design goals to be traded off against each other. In addition, different concepts and interpretations of trust based on different modelling theories and principles can be accommodated and tailored to suit specific application contexts and situations.

The emphasis of the approach is to facilitate the consideration of trust early in the design process, so that trust considerations will be used as a driving force, among others, to guide the exploration and evaluation of potential design alternatives.

Section 2 presents an overview of the $i*$ framework, introducing its basic concepts using a smart card example. Section 3 considers the modelling of trust for system design. Section 4 analyzes deception, fraud and trust in smart card systems. An outline of the qualitative evaluation method for propagating satisficing judgements across the network model is also provided. Section 5 studies the impact of trust on system design, illustrating that different kinds or degrees of trust maybe needed for different system configurations. Section 6 discusses related work. Only a subset of the features of $i*$ are illustrated in this paper.

2   An Overview of the *i* Framework

The framework includes a Strategic Dependency model – for describing the network of relationships among actors, and a Strategic Rationale model – for describing and supporting the reasoning that each actor has about its relationships with other actors. The basic concepts and examples of SD and SR model are introduced in section 2.1 and 2.3 respectively.

2.1   The Basic Strategic Dependency Model

A Strategic Dependency (SD) model consists of a set of nodes and links. Each node represents an actor, and each link between two actors indicates that one actor depends on the other for something in order that the former may attain some goal. We call the depending actor the depender, and the actor who is depended upon the dependee. The object around which the dependency relationship centers is called the dependum. By depending on another actor for a dependum, an actor (the depender) is able to achieve goals that it was not able to without the dependency, or not as easily or as well. At the same time, the depender becomes vulnerable. If the dependee fails to deliver the dependum, the depender would be adversely affected in its ability to achieve its goals.

Fig. 1 shows a Strategic Dependency model for a generic smart card-based payment system. There are six actors in the system, who are connected by various types of dependency links to form a social network. A cardholder depends on a card issuer to be allocated a smart card, for the terminal owner depends on him to present his card for each transaction. The card issuer in turn depends on the card manufacturer and software manufacturer to provide cards, devices, and software. The data owner is the one who has control of the data within the card. He depends on the terminal owner to submit transaction information to the central database.

The Strategic Dependency model distinguishes among several types of dependencies, based on the ontological category of the dependum. In a *goal dependency*, an actor depends on another to make a condition in the world come true. Because only an end state or outcome is specified, the dependee is given the freedom to choose how to achieve it. In the example of Fig. 1, the goal dependency "new account be created" from the card issuer to the data owner means that it is up to the data owner to decide how to create a new account. The card issuer does not care how a new account is created, what matters is that, for each card, an account should be created.

In a *task dependency*, an actor depends on another to perform an activity. The depender's goal for having the activity performed is not given. The activity description specifies a particular course of action. The card issuer depends on the cardholder to apply for a card via a task dependency by specifying standard application procedures. If the card issuer were to indicate the steps for the data owner to create a new account, then the data owner would be related to the card issuer by a task dependency.

In a *resource dependency*, an actor depends on another for the availability of an entity. The depender takes the availability of the resource to be unproblematic. In Fig. 1, the card issuer's dependencies on the card manufacturer for cards and devices, the manufacturers' dependencies on card issuer for payment are modelled as resource dependencies.

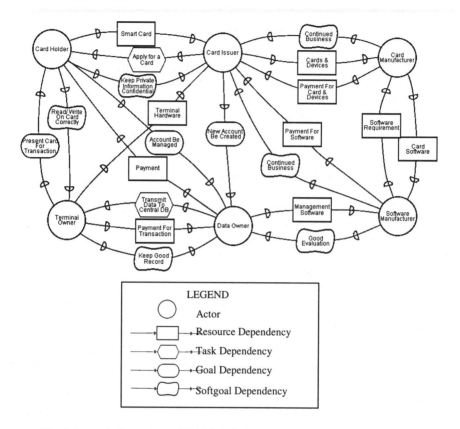

Fig. 1. Strategic Dependency (SD) Model of Smart Casr Based Payment System

The fourth type of dependency, *softgoal dependency*, is a variant of the first. It is different from a (hard) goal dependency in that there are no *a priori*, cut-and-dry criteria for what constitutes meeting the goal. The meaning of a softgoal is specified in terms of the methods that are chosen in the course of pursuing the goal. The dependee contributes to the identification of alternatives, but the decision is taken by the depender. The notion of the softgoal allows the model to deal with many of the usually informal concepts. For example, the manufacturers' dependencies on the card issuer for continued business can be achieved in different ways. The desired style of continued business is ultimately decided by the depender. The cardholder's softgoal dependency on the card issuer to "keep private information confidential" indicates that there is not a clear-cut criterion for the achievement of confidentiality. The four types of dependencies reflect different types of freedom that is allowed in the relationship between depender and dependee.

The Strategic Dependency model of Fig. 1 is not meant to be a complete and accurate description of any particular smart card system. It is intended only for illustrating the features of *i**.

2.2   Roles, Positions, and Agents

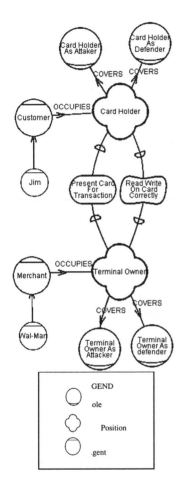

Fig. 2. Strategic Dependency model with roles, positions, and agents

In *i**, the term *actor* is used to refer generically to any unit to which intentional dependencies can be ascribed. To model complex relationships among social actors, we further define the concepts of agents, roles, and positions, each of which is an actor in a more specialized sense. A basic Strategic Dependency model can be extended by refining the notion of actor into notions of role, position, and agent.

An agent is an actor with concrete, physical manifestations, such as a human individual. An agent has dependencies that apply regardless of what role he/she/it happens to be playing. For example, if Jim, a cardholder desires a good credit record, he actually wants the credit record to go towards his personal self, not to the positions and abstract roles that Jim might occupy or play. We use the term agent instead of person for generality, so that it can be used to refer to human as well as artificial (hardware, software, or organizational) agents. In Fig. 2, customer and merchant are represented as agents.

A *role* is an abstract characterization of the behavior of a social actor within some specialized context or domain of endeavor. Dependencies are associated with a role when these dependencies apply regardless of who plays the role. For example, we consider attacker and defender as two roles any actor can play. No matter who plays the role of attacker, he will have a high level goal of "attack". Regardless of who plays the role of defender, he will have the goal of "defense".

A *position* is intermediate in abstraction between a role and an agent. It is a set of roles typically played by one agent. Positions can *cover* roles, agents can *occupy* positions, and agents can also *play* roles directly. Fig. 2 shows a fragment of the Strategic Dependency model from the smart card example with agents, roles, and positions. In this partial model, a cardholder position covers two roles of cardholder as attacker and cardholder as defender. The position of cardholder is occupied by the agent "customer". The "INS" construct represents the instance-and-class relation. For example, Wal-Mart is an instance of merchant, and Jim is an instance of customer. The "ISA" construct expresses conceptual generalization/ specialization. For example, a bank is a kind of financial institution. These constructs are used to simplify the presentation of strategic models with roles, positions, and agents.

There can be dependencies from an agent to the position it occupies. For example, a merchant who occupies the position of terminal owner depends on that position to attract more customers. Otherwise, he may choose not to occupy that position.

Roles, positions, and agents can each have subparts. Aggregate actors are not compositional with respect to intention. Each actor, regardless of whether it has parts, or is part of a larger whole, is taken to be intentional. Each actor has inherent freedom and is therefore ultimately unpredictable. There can be intentional dependencies between the whole and its parts, e.g., a dependency by the whole on its parts to maintain unity.

In *i**, the distinction among agents, roles, and positions can be used to model different kinds of trust. For example, a customer may trust a bank, without necessarily trusting specific employees in the bank. The bank as an abstract concept is modelled in terms of roles and positions, with checks and balances among them, and protection and control mechanisms. Conversely, there can be personal and goodwill-based trust on the bank employees (modelled as agents), even when the customer doubts the adequacy of the bank's protection mechanisms.

2.3   The Strategic Rationale Model

The Strategic Rationale (SR) model of *i** provides a more detailed level of modelling by looking "inside" actors to model internal intentional relationships. Intentional elements (goals, tasks, resources, and softgoals) appear in SR models not only as external dependencies, but also as internal elements arranged into a hierarchy of means-ends, task-decompositions and contribution relationships. The SR model in Fig. 3 elaborates on the rationale of a smart card manufacturer. From the top right corner of the model in Fig. 3, we can see that the card manufacturer's business goal is to "Manufacture Card Hardware". Two different means for accomplishing this goal are to "Provide Total Card Solution" (such as the Mondex solution [16]), and to "Provide Simple Card Solution" (such as the Millicent Solution [16]). They are connected to the goal with means-ends links. "Provide Total Card Solution" will help the security of the system (represented with a "Help" contribution link to "Security"), while "Provide Simple Card Solution" does not affect the security of the system too much only if it is applied to small amount card. The Simple Card Solution is good for the goal of "Low Cost" whereas the Total Card Solution is bad.

The task "Provide Total Card Solution" can be decomposed into three sub-components (connected with task-decomposition links): the sub-task of "Develop Card Solution" and "Manufacture Card & Devices", and the sub-goal of "Get Paid". The decomposition relationship in *i** concerns more on the strategic and intentional sub-components of a task than the sub-procedure or concrete data components.

The positive contribution types for softgoals are Help (positive but not by itself sufficient to meet the higher goal), Make (positive & sufficient) and Some+ (positive in unknown degree). The corresponding negative types are Hurt, Break and Some-. And means if all subgoals are met, then the higher goal will be sufficiently met. Or means the higher goal will be sufficiently met if any of its subgoals are met. During system analysis and design, softgoals such as Low Cost and Security are systematically refined until they can be operationalized and implemented [11,12]. Unlike functional

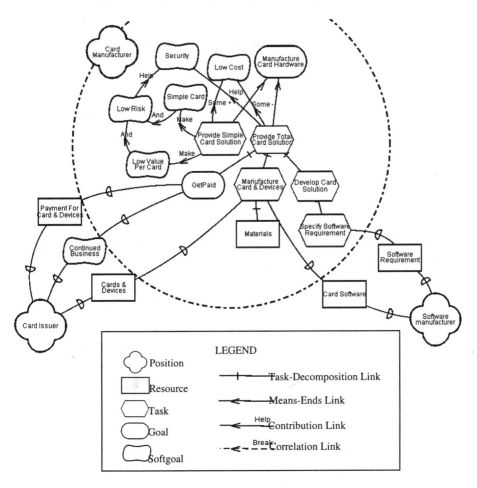

Fig. 3. Strategic Rationale (SR) Model of Card Manufacturer

goals, non-functional qualities represented as softgoals frequently interact or interfere with each other, so the graph of contributions is usually not a strict tree structure [7].

SR models, as exemplified by Fig. 3, will be used to explore and reason about alternatives and their contributions to goals including trust, which helps to lead their satisfaction.

3   Modelling Trust in *i** for System Design

The intentional and strategic underpinnings of *i** concepts lend themselves well to the modelling of trust. In the SD model, notions of opportunity and vulnerability are already built into the central modelling construct of the intentional dependency. In considering a network of dependencies as a potential configuration to adopt, the designer is led to question the viability of each dependency. Whether the depender trusts the dependee to deliver the dependum is often a part of that assessment. In this section, we illustrate how trust can be explicitly modelled and reasoned about with *i**. Trust would be treated together with other design issues and considerations. The network of intentional relationships among strategic actors provides a structural representation of potential design configurations under consideration.

Trustworthiness (of a dependee from the viewpoint of a depender) is treated as a design goal. During design, the designer would look for configurations of agents and relationships that would meet the design goals.

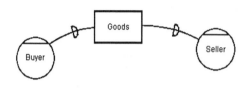

Fig.4(a) A Single Resource Dependency

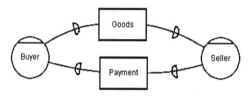

Fig.4(b). An Enforcement

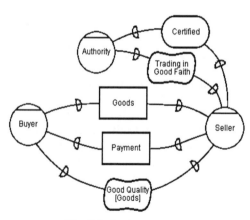

Fig.4(c). An Assurance

Trust is modelled through softgoals so that different concepts of trust can be accommodated, by refining it into constituent concepts appropriate for the design context. The softgoal qualitative reasoning approach also allows trust related goals to be traded off with other design goals to achieve overall design objectives.

Consider a simple example in which a buyer depends on goods from a seller. For simplicity, let us treat this as a single resource dependency (as in Fig. 4(a)). The depender is concerned about the viability of this relationship. The buyer's trust in the seller is one factor, but there are also other structural relationships that can help in the judgement. For example, if the seller also has a dependency on the buyer – such as the dependency on payment (as in Fig. 4(b)), then the viability of the «Goods» dependency would be *enforceable*, since the buyer can withhold payment. (A more detailed analysis would include temporal relationships such as precedence and whether the dependencies are long-term.)

Another configuration may include a third party authority, which provides *assurance* about the quality of the goods, for example, through certification as a control (see Fig. 4(c)).

These trust related factors that contribute to the viability of a design configuration can be modelled in a SR model (as in Fig. 5) so that their overall effect and interactions can be assessed together. In this case, trustworthiness is modelled as a softgoal contributing to the dependee's judgement of the viability of buying goods from the seller (in this

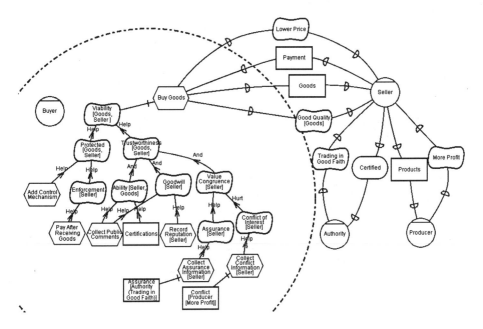

Fig. 5. A SR Model of Buyer and its trust factors

fashion). For the seller to be considered trustworthy, the buyer wants the seller to have the ability to provide the goods, goodwill, and congruent interest and values. The goal of Value Congruence is supported by information from some authority, which provides *assurance*. Conversely, a conflict of interest would hurt the value congruence goal, making the seller less trustworthy.

The refinement of trustworthiness softgoal leads to the requirements for public comment collection mechanism, reputation recording mechanism, assurance and conflict information collecting mechanism, etc. These requirements will be embodied in the design of the final system.

Apart from trustworthiness, there are control mechanisms that also contribute to viability. The separate «Protected» softgoal captures the intuition that the stronger the protection mechanisms, the less need there is for trust. The combination of the two softgoals lead to a judgement of viability.

So far, we have only considered trust from the trustor's viewpoint. This would be enough if a trustee does not care whether the trustor trusts her or not. However, if the trustee wants to be trusted, she might take deliberate action to increase her own trustworthiness in eyes of the trustor. This may be modelled as a softgoal "Trusting" in the trustee. If the trustee's knowledge on the trustor's propensity of trust tells her that she is not likely to win his trusts, she might turn to a more risk-taking trustor to cooperate.

This may be used to model deception. In Fig. 6, Buyer wants Seller to trust him on the payment. If he wants to cheat, one of the efforts he can make is to look for a more gullible seller. This Seller is chosen as the Buyer knows that he has no requirement on a guarantor. The Buyer may also deceive by falsifying a credit record. This model shows

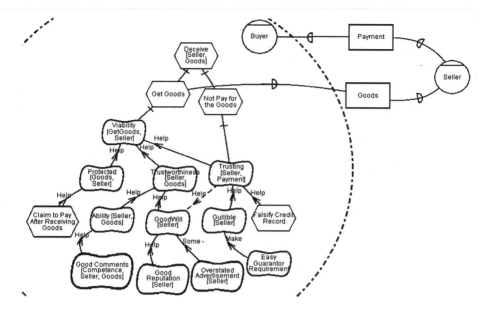

Fig. 6. Trust and Deception

that the depender's trust actually brings vulnerabilities, as it increases the opportunities of fraud and deception, which takes the advantage of the trustor's willingness to assume risk.

4 Deception, Fraud and Trust in the Smart Card Example

In this section, we apply the trust modelling to the smart card example.

4.1 Exploring Design Alternatives

During system design, designers address trust issues by considering how trustworthy the future system should be, who should be trustworthy in the system, what design alternatives will make the system more trustworthy, etc. For example, a system that requires a high trust level may lead to a design with built-in reputation mechanisms. Reputation mechanisms can be distributed into each agent, or be centralized into a trusted third party according to the demands of agents and other design goals, such as performance, flexibility, cost, etc. Reputation can be accumulated through either experimental conversations or real cooperation experiences depending on whether the trust is focusing on information or on behaviors.

In Fig. 7, different design alternatives in stored value smart card systems are explored. To establish a smart card system, several goals and softgoals need to be achieved. For example, in order to achieve the goal of Card Device Be Designed, a system may either use reloadable card or use one-off card. The two alternative work styles of point of sale (POS) may be either online or offline. Combinations of different choices among

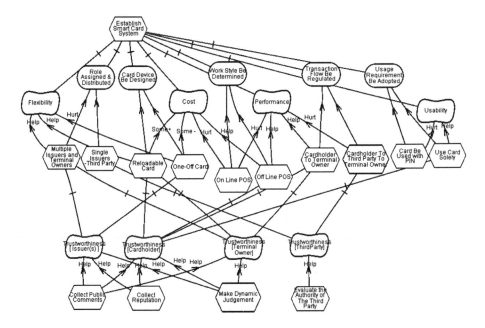

Fig. 7. Design Alternatives in Stored Value Smart Card System

the alternatives result in different design configurations for the future system. The non-functional qualities of the design are determined by what alternatives have been chosen. For example, Online POS reduces the performance of the system, as data has to be transmitted for each transaction. It also increases the cost of the system for the need of the data transmission devices and software. However, Online POS does not have trust requirement, while offline POS requires the cardholder to be trustworthy because the chip embedded in the card contains all the information necessary to identify the card and its value. If the card is using a bogus chip, or a valid chip with falsely added values, the terminal owner or bank will be hurt. Driven by the trustworthiness requirements, certain reputation or comment collection mechanism will be considered for incorporation in the system design.

4.2 Assessing the Potential Risks of Frauds and Attacks in the System

For each design configuration resulting from an initial trade-off of functional and non-functional requirements, we still need to assess its viability due to the possibility of potential attacks. Generally, a dependee attacks or defrauds a depender by taking advantage of the dependency and trust. Whenever a potential attack is judged to be strong enough to make a dependency unviable, the system will need either additional trust, or further protection mechanisms to become viable.

As shown in the Strategic Rationale model of Fig. 8, the cardholder depends on the terminal owner to «Read/Write Card Correctly». To address the vulnerability arising from this dependency, we consider the case where the terminal owner is not trustworthy.

To do this, we elaborate on the role of Terminal Owner As Attacker. There are a number of potential attacks that are sufficient to make this dependency not viable (**Break**).

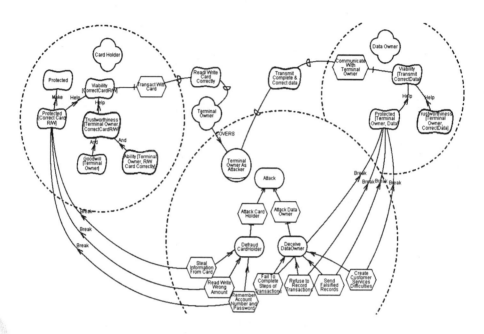

Fig. 8. Assesing the Potential Fraud and Attack Violating Trust

In Fig. 8, we model potential attacks (including fraud) as negative contributions from the attackers (from their specific methods of attack) toward the Protected softgoal of dependers. A **Break** contribution indicates that the attack is sufficient to make the softgoal unviable. For clarity of analysis, we place the attack-related intentional elements of the card issuer into a role called «Terminal Owner As Attacker». Details of the attack methods (e.g., Steal Card Information, Send Falsified Records) can be elaborated by further decomposition and means-ends analysis. Other internal details of the Terminal Owner are not relevant and are thus not shown. Negative contribution links are used to show attacks on more specific vulnerabilities of the depender (e.g., refinements of «Transact with Card»). The refinements (and possible attackroutes) may be based on analysis of the SD and SR models of the normal operations of the smart card, e.g., what resources an actor accesses, what types of interactions exist, etc.

With the knowledge of some potential attacks and frauds, dependers may first look for trustworthy partners, or change their methods of operation, or add control mechanisms (countermeasures) to protect their interests. A countermeasure may prevent the attack from happening by either making it technically impossible, or by eliminating the attacker's intent of attack.

Fig. 9 shows a SR model with defensive actions as well as attacks. Protection mechanisms are adopted to counteract specific attacks. In some cases, the protections are

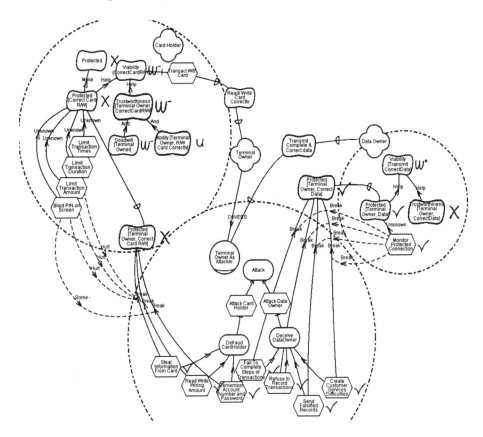

Fig. 9. Qualitative Reasoning on Protections to Potential Attacks Regaining Trust

sufficient to defeat a strong attack (defense **Break** link (dotted arrow) pointing to an attack Break link). In other cases, countermeasures are only partially effective in defending against their respective attacks (through the **Hurt** or **Some-** contribution types). In Fig. 9, the Data Owner's defenses are sufficient, while the cardholder's measures are only partially effective.

4.3   Qualitative Reasoning

The qualitative reasoning process used is an interactive labeling algorithm, which propagates a series of labels through the modelling framework. A label (or satisficing status) on a node is used to indicate whether that intentional element (goal, task, resource, or softgoal) is viable or not (e.g., whether a softgoal is sufficiently met). A qualitative reasoning scheme is used. Labels can have values such as Satisfied «√», Denied «×», Weakly Satisfied «W⁺» and Weakly Denied «W⁻», Undecided «U», etc. Leaf nodes (those with no incoming contributions) are given labels by the analyst based on judgement of their independent viability. These values are then propagated «upwards» through the contribution network. The viability of the overall system appears in the high level nodes of the various stakeholders. It is an interactive one requiring the analyst to make judgements whenever the outcome is inconclusive given the combination of incoming contributions.

In Fig. 9, the analyst labels all the attack leaf nodes as "Satisficed" since they are all judged to be possible. Similarly all the defense leaf nodes are judged to be viable, thus labelled "Satisificed". The values are then propagated along contribution links. Before the adding of defense nodes, Protected softgoal was labelled as Denied, because of the potential strong attacks from Terminal Owner. However, as countermeasures are added into system configuration, the influences of the attacks will be weakened.

Regarding the cardholder's dependency on the terminal owner for "Read/ Write Card Correctly", there are three possible attacks. One of them "Steal Card Info" is counteracted by three defense measures, though each one is partial (Hurt). Another attack "Remember Account Number & Password" has a defense of unknown strength (Some-). The third attack has no defensive measure. The softgoal "Protected [Correct Card R/W]" is thus judged to be weakly unviable (W^-). As the Trustworthiness softgoal is also weakly denied, the Viability softgoal is Denied ($\times$).

On the other side, as the Data Owner's protection mechanism could sufficiently defeat the four possible attacks, the "Protected [Terminal Owner, Correct Data]" softgoal is thus judged to be viable ($\sqrt{}$). Though the Trustworthiness of Terminal Owner is also labelled as Weakly Denied (W^-), the Viability is judged as Weakly Satisfied (W^+).

Untrustworthiness and potential attacks lead to the erosion of viability of the smart card system. Incorporating sufficient countermeasures or trustworthiness evidence restores viability.

5   The Impacts of Trust Relationships on System Configuration

In the above modelling, the various participants in a smart card system were modelled as positions and analyzed generally. The trust requirements are considered mainly in the level of abstract roles, which focuses more on the control/protection facets of trust. However, in real world smart card systems, specific organizational parties or personnels occupy these positions. Thus, to actually understand their trust situations, we have to apply the generic model to the real world configurations. We consider two representative kind of smart card based systems. One is the Digital Stored Value Card, the other is the Prepaid Phone Card [17]. The six abstract positions in the generic smart card system (Fig. 1) would now be mapped into concrete real world agents.

5.1   Digital Stored Value Cards

These are payment cards intended to be substitutes for cash. Both Mondex and VisaCash are examples of this type of system. Cardholders are customers. Terminal owners are merchants. Data owner and card issuer are both the financial institutions that support the system. Card manufacturer and software manufacturer are both technology providers like Mondex.

In such a configuration, the previously isolated positions of data owner and card issuer are occupied by the same physical agent, namely, Financial Institution. Similarly, card manufacturer and software manufacturer are combined into one physical agent – the Smart Card Technology Provider. Fig. 10 describes the Strategic Dependency model of a digital stored value card. Here the software manufacturer's attack to card manufacturer can be ignored since they belong to the same agent – the smart card technology company. These two pair of positions won't have any problem to trust each other as the intent of deception and fraud disappears. And the costs on control mechanisms could also be saved. (The dotted circles are used here to highlight the area of interest to the reader. They are not part of the notation.)

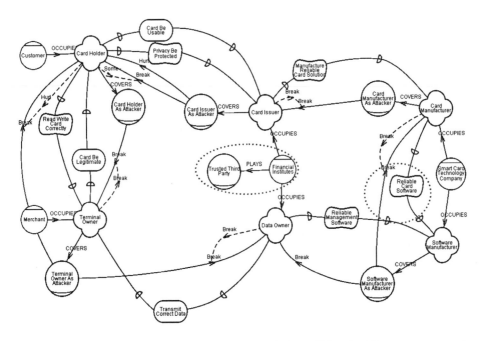

Fig. 10. A Strategic Dependency model of a Store Value Card System

After several such card systems are established, customers have to decide which financial institute and merchants are trustworthier, and which card system to enter. When a customer feels that there are negative symptoms for the trustworthiness of his current card issuer or terminal owner, he might cancel his account from that system, and look for a trustworthier one. As digital stored value cards may involve considerable amount of money, all the three agents in the system, the financial institute, the merchant, and the customer should be cautious on their decision making process. That is, they should set a higher threshold of trust, and should use policies such as "not to trust unless some positive hints or symptoms were detected." Moreover, as the Financial Institution has more authorities comparing to the other two, it can act as a trusted third party in the transactions. For example, money is first withdrawn from the customer's account to the bank, and is forwarded by the bank to the merchant instead of transmitted from the customer to the merchant directly.

5.2   Prepaid Phone Cards

These are simply special-use stored value cards. Cardholders are the customers. Terminal owner, data owner, manufacturer and card issuer are all combined into one agent – the phone company. Fig. 11 shows the Strategic Dependency model of a prepaid card system. Under such a system configuration, more attack-defense pairs (in dotted circles) will disappear, and only the goodwillness of two physical agents needs to be judged. Only four possible attacks need to be considered now. Three of them are from the Phone Company, which are to hurt privacy, to issue unusable card, to read and write card incorrectly. The other attack is from the cardholder, who might use an illegitimate card.

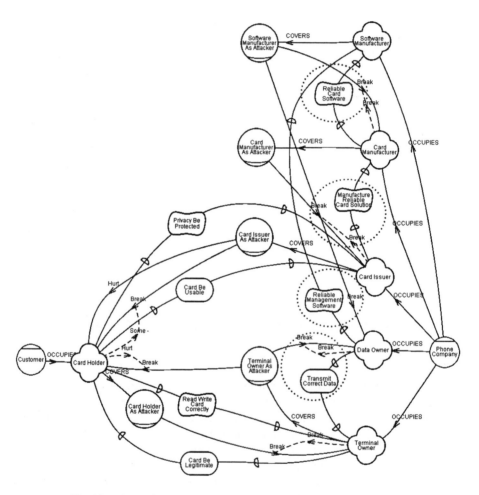

Fig. 11. A Strategic Dependency Model of Prepaid Phone Card System

As a prepaid phone card involves comparatively small amount of money, both the phone company and the customer tend to be relaxed on making trust decisions. They might use policies of "to trust unless there is bad feedback or unmet expectations." Typically, no specific trust-related mechanisms would be designed in due to their costs.

Note that each time new positions are created, additional trust relations need to be evaluated, and the possibility of new attacks arises. These models reflect Schneier's observation that the fewer splits one makes, the more trustable the target system might be [17].

6 Discussion

Trust is becoming an increasingly important issue in the design of many kinds of information systems. Many new kinds of technologies are being used in new contexts and socio-technical configurations that have not been tried before. Therefore, Assessing and establishing trust is an important part of the design process. The uncertainties and concerns of various stakeholders and participants need to be considered and addressed.

As trust is traditionally associated with human agents, it is difficult to deal with within a technical systems design framework. Recent research on trust in agent societies has examined and elaborated on the nature of trust (e.g., [10] [4] [22]). The contributing factors to trust, the characteristics of trustor and trustee, and the relationships between trust and other concepts have been explored. Some approaches adopt formal or quantitative evaluation mechanisms of trustworthiness, such as reputation systems [28], observation models [15], and so on. Others contribute to the solving of particular trust issues in e-commerce, e.g., how to achieve both trust and privacy in an anonymous environment [19], what makes a web page more trustworthy [20], etc.

In this paper, we have proposed a design framework that can potentially incorporate ideas and techniques from the above diverse approaches to trust. In the *i** approach, trust is not treated as a distinguished concept with special semantics. Instead, trust is treated as a non-functional requirement that arises in multi-agent collaborative configurations. In these configurations agents depend on each other in order to succeed. Some configurations may require more trust than others. For example, the financial institution and the customer in a stored valued card system are taking more risks than the phone company and customer in a prepaid phone card system. Each configuration is predicated on certain kinds of trust by some actor (roles or position or embodied agents) on some other actors. For example, in a smart card system, there are personal goodwill-based trust, as well as control and protection. Section 4 discussed mainly control and protection-based trust, while in section 5, at the physical agent level, we considers more about the personal goodwill based trust.

*i** offers a structural representation that shows relationships among actors with which trust issues may arise. During early design stages, non-functional requirements, modelled as "softgoals", are analyzed and evaluated through qualitative reasoning. Softgoals and other design goals are systematically refined and reduced until they are operationalized. Therefore, trust considerations are taken into account when other design goals (such as functionality, usability, performance, costs and time-to-market) are being addressed.

For example, when designing an e-commerce system, one may discover that one of the potential designs (a configuration of humans, software agents, and infrastructure) is attractive in costs, time-to-market, and other attributes, but require a higher degree of

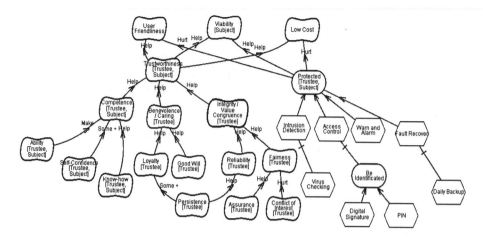

Fig. 12. Interrelationschip of Trust and Other Design Goals

trust than available. The design would then use the combination of design goals to guide the further exploration of the space of design alternatives, until the trust goals are also met.

In this paper, we introduced the approach with a simple buyer-seller example, and further illustrated it in a more concrete setting with the smart card example. In an actual application, one could expect the analysis to be more elaborate, taking more relevant trust factors into account. Fig. 12 shows a few more contributing factors to trust discussed in the literature. It also shows their interactions with other design factors such as usability and cost.

The *i** approach encourages and facilitates the analysis of trust-related issues within the full operational and social context of the involved actors. The models can be used to encompass normal-case operational procedures, potential attacks, countermeasures against perceived threats, as well as factors not directly related to trust. In the smart card example, trust issues can be examined in the context of customers' options among different kinds of payment systems, e.g., credit card, cash, smart card, with tradeoffs among convenience, security, and privacy issues. As trust is not a hard-wired concept in *i** framework, it is flexible enough to handle the different notions and meaning of trust that may apply to a certain context or problem domain. It is also capable of tackling the different issues related to trust.

This approach is complementary to and benefits from the various theories and techniques currently being developed for specifically addressing trust (e.g., [4] [5] [22]). *i** offers a structural representation of intentional relationships among actors and within actors, as well as structural concepts such as intentional agents, roles, and positions. These concepts provide a structural framework for integrating other concepts and techniques for dealing with trust. For example, the qualitative reasoning approach of *i** may be used in a first-pass preliminary analysis, to be followed by techniques with stronger semantics e.g., reputation mechanisms for dynamic evaluation of trust [28], probabilistic estimation of trust levels [2], analysis using game theory [1] and formal logic [8, 21].

The framework has been used to analyze requirements in several application domains such as smart card systems, online travel business, healthcare information systems, and telecommunication software systems. Its application to the analysis of trust in the design of real systems remains to be tested.

In future work, we plan to use *i** to analyze broader issues related to trust, including issues of privacy, power, and public policy, as discussed, for example in [20]. A prototype tool has been developed and it supports modelling and reasoning based on the *i** framework.

References

1. Birk, A.: Trust in an N-Player Iterated Prisoner's Dilemma. Proc. of Deception, Fraud and Trust in Agent Societies Workshop, Autonomous Agent 1999, Seattle, USA, May 1999.
2. Biswas, A., Mundhe, M., Debnath, S., Sen, S.: A Bayesian Network Based Approach for Modeling Agent Relationships. Proc. of Deception, Fraud and Trust in Agent Societies Workshop, Autonomous Agent 1999, Seattle, USA, May 1999.
3. Camp, L.J.: Trust and Risk in Internet Commerce. MIT Press, 2000.
4. Castelfranchi, C. and Falcone, R.: The Dynamics of Trust: From Beliefs to Action. Proc. of Deception, Fraud and Trust in Agent Societies Workshop, Autonomous Agent 1999, Seattle, USA, May 1999. 49-60.
5. Castelfranchi, C. and Falcone, R.: Does Control Reduce or Increase Trust? A complex relationship. Proc. of Deception, Fraud and Trust in Agent Societies workshop, Autonomous Agent 2000, Barcelona, Spain, June 2000. 49-60.
6. Chung, L.: Representing and Using Non-Functional Requirements for the Information System Development: A Process-Oriented Approach, Ph.D. Thesis, also Tech. Report DKBS-TR-93-1, Dept. of Computer Science, University of Toronto, June 1993.
7. Chung, L., Nixon, B. A., Yu, E., Mylopoulos, J.: Non-Functional Requirements in Software Engineering. Kluwer Academic Publishers, 2000.
8. Jonker, C. M., Treur, J.: Formal Analysis of Models for the Dynamics of Trust Based on Experiences. Proc. of Deception, Fraud and Trust in Agent Societies Workshop, Autonomous Agent 1999, Seattle, USA, May 1999.
9. Marsh, S., Meech, J.F., Dabbour, A.: Putting Trust into E-Commerce – One Page at a Time. Proc. of Deception, Fraud and Trust in Agent Societies Workshop, Autonomous Agents'2000. Barcelona, Spain, June 2000.
10. Mayer, R. C., Davis, J. H., Schoorman, F. D.: An Integration Model of Organizational Trust, Academy of Management. The Academy of Management Review, Vol. 20(3), Jul 1995.
11. Mylopoulos, J., Chung, L., Wang, H.Q., Liao, S., Yu, E.: Extending Object-Oriented Analysis to Explore Alternatives. IEEE Software. January/February 2001.
12. Mylopoulos, J., Chung, L., Nixon, B.: Representing and Using the Non-Functional Requirements: A Process-Oriented Approach, IEEE Trans. Software Engineering. 18(6), June 1992.
13. Mylopoulos, J., Chung, L., Yu, E.: From Object-Oriented to Goal-Oriented Requirements Analysis, Communications of the ACM, 42(1): 31-37, January 1999.

14. Salter, S., Saydjari, O., Schneier, B., Wallner, J.: Toward a Secure System Engineering Methodology, Proc. of New Security Paradigms Workshop 1998, IEEE Computer Society Press.
15. Schillo, M., Funk, P., Rovatsos, M.: Who Can You Trust: Dealing with Deception. Proc. of Deception, Fraud and Trust in Agent Societies Workshop, Autonomous Agent 1999, Seattle, USA, May 1999.
16. Schneider, F. B.: Trust in Cyberspace. Committee on Information Systems Trustworthiness, Computer Science and Telecommunications Board, National Research Council. Washington, D.C.: National Academy Press, 1999. Also available at http://cryptome.org/tic.htm.
17. Schneier, B.: Attack Trees Modelling Security Threats. Dr.Dobb's Journal, December 1999. Also available at http://www.counterpane.com/attacktrees-ddj-ft.html.
18. Schneier, B., Shostack, A.: Breaking Up Is Hard To Do: Modelling Security Threats for Smart Cards. Available at http://www.counterpane.com/smart-card-threats.html. Also First USENIX Symposium on Smart Cards, USENIX Press, to appear.
19. Serban, R, van de Riet, R.: Hiding in a Group: Who is Responsible? Proc. of Deception, Fraud and Trust in Agent Societies Workshop, Autonomous Agents'2000. Barcelona, Spain, June, 2000.
20. Stalder, F., Clement, A.: Exploring Policy Issues of Electronic Cash: The Mondex Case, Canadian Journal of Communication, 24(2). 1999.
21. Tan, Y., Thoen, W.: Formal Aspects of a Generic Model of Trust for Electronic Commerce. Proc. of Deception, Fraud and Trust in Agent Societies Workshop, Autonomous Agents'2000. Barcelona, Spain, June 2000.
22. Tan, Y., Thoen, W.: Towards a Generic Model of Trust for Electronic Commerce. Proc. of Deception, Fraud and Trust in Agent Societies Workshop, Autonomous Agent 1999, Seattle, USA, May 1999.
23. Yu, E.: Modelling Strategic Relationships for Process Reengineering, Ph.D. thesis, also Tech. Report DKBS-TR-94-6, Dept. of Computer Science, University of Toronto, 1995.
24. Yu, E.: Towards Modelling and Reasoning Support for Early-Phase Requirements Engineering, Proc. 3rd IEEE Int. Symp. On Requirements Engineering (RE'97), Annapolis, Maryland, USA, January 1997.
25. Yu, E., Mylopoulos, J.: From E-R to 'A-R' – Modelling Strategic Relationships for Business Process Reengineering, Int. Journal of Intelligent and Cooperative Information Systems, 4(2&3), 1995, pp.125-144.
26. Yu, E., Mylopoulos, J.: Understanding 'Why' in Software Process Modelling, Analysis, and Design, Proceedings of 16th International Conference On Software Engineering, May 1994, pp. 159-168.
27. Yu, E., Mylopoulos, J., Lespérance, Y.: AI Models for Business Process Reengineering, IEEE Expert, August 1996, pp.16-23.
28. Zacharia, G.: Trust Management through Reputation Mechanisms. Proc. of Deception, Fraud and Trust in Agent Societies Workshop, Autonomous Agent 1999, Seattle, USA, May 1999.

Author Index

Lecture Notes in Artificial Intelligence (LNAI)

Lecture Notes in Computer Science